FISH, FISHING, AND FISHERIES

FISHERY MANAGEMENT

FISH, FISHING, AND FISHERIES

Additional books in this series can be found on Nova's website
under the Series tab.

Additional E-books in this series can be found on Nova's website
under the E-books tab.

FISH, FISHING, AND FISHERIES

FISHERY MANAGEMENT

JANICE S. INTILLI
EDITOR

Nova Science Publishers, Inc.
New York

Library of Congress Cataloging-in-Publication Data

Fishery management / editors, Janice S. Intilli.
p. cm.
Includes index.
ISBN 978-1-61209-682-7 (hardcover)
1. Fishery management. I. Intilli, Janice S.
SH328.F575 2011
338.3'727--dc22

2011001546

Published by Nova Science Publishers, Inc. † New York

CONTENTS

PREFACE

Fisheries management draws on fisheries science in order to find ways to protect fishery resources so sustainable exploitation is possible. This new book presents topical research in the study of fishery management including coral reef fisheries under threat from overfishing; implementing population viability analysis into fisheries management; machine learning analysis for fisheries research; the marine fisheries policy in Nigeria and evaluating the recovery plan for European hake fishery

Chapter 1- Coral reefs are one of the world's most diverse and productive ecosystems, but are seriously under threat from the problems caused by overfishing. Recent population booms, combined with a low level of economic development make many tropical coastal communities highly dependent on natural resources, leading to extraction rates often significantly exceeding maximum sustainable yield. A lack of suitable alternatives, in terms of both livelihoods and food security, means the use of highly impacting gear types persists, a scenario characteristic of Malthusian overfishing. These economic and socio-biological concerns, typically associated with small-scale artisanal fisheries as seen around coral reefs, bring with them a unique set of problems for conservation managers to overcome. Traditionally in tropical communities, a strict approach of top-down management has been employed, focusing on command and control strategies, but these have tended to fail due, amongst other factors, to a lack of compliance from local stakeholders. Recently, however, there has been increased emphasis on adaptive community-based management, which acts both to empower local user groups, and to build upon the wealth of local ecological knowledge traditional fishing families possess. This chapter aims to introduce coral reef fisheries, describe the distinctive dynamics of coral reef artisanal fisheries, and explore the idea behind community management.

Chapter 2- This paper focuses on the regulatory tools and related mesaures (e.g., acts, rules, policies and plans) of Bangladesh that define the management and conservation of inland capture fisheries in the rivers, floodplains and *beel lakes* of Bangladesh. The goal of this paper is to help policy-makers and scientific communities to rethink conservation policies and help them realize that in implementing and developing policy local level participation in decision-making and their livelihoods issues are vital. This study recognizes that the natural resource policy tools of Bangladesh are very broad and have long history of managing natural resources but with poor enforcement. Fisheries management also lacks participatory dialogue during the policy development process. The policy tools, therefore, are not very effective and so lead to poor compliance and impede the sustainable development of inland fisheries. Given the apparent inadequacy of the existing management and regulatory tools, this paper argues for multi-stakeholder decision making involving fishers and other local stakeholders at the policy formation stage which can lead to better management of inland capture fisheries.

Chapter 3- Fish populations worldwide have experienced severe decline, and some of the species are facing a serious threat of extinction. Population simulations are currently the most promising approach for the assessment of extinction risk and the evaluation of alternative management measures. The population viability analysis (PVA) represents a method that is able to cope with uncertainties present in ecological data and to support decision-making process within the field of fishery management. This chapter presents an overview of the development of the PVA within the field of fishery science, along with an evaluation of the success of its application in fishery management, and major challenges in its future implementation. Potential applications of PVA are very diverse and, beside the estimation of sustainable fishery levels, they were used to address a number of different issues, such as the efficiency of the minimum length limits, extinction risk of the target and bycatch species, impact of the climate change, effects of habitat loss and degradation, and other. Despite some unresolved limitations and problems in the PVA development and implementation, scientific knowledge and fishery management have both earned immensely from this process. PVA should be recognized as an important tool for the process of research and management planning and policy development, and as an integral evaluation and planning component within the adaptive management process. If used with necessary caution, PVA can provide invaluable help in current worldwide efforts for achieving sustainable fishery.

Chapter 4- Authors show how to apply Classification and Regression Trees to deal with problems in fisheries research. Classification and Regression Trees are techniques of Machine Learning that allow the recognition of complex patterns in data. This permits the extraction of hidden information contained in large databases, which represents a major challenge in fisheries research. Authors illustrate they explanation with a non-published case of study as well as leaning on another example already published by the authors, although very different kinds of study can be treated with these methods. The application process is made up of three main stages. The result of a first data pre-processing phase is the training data set. The second stage involves the application of one or several induction algorithms, which translates into a set of trees. The analysis and interpretation of these trees, in the third and last stage, allows the generation of new knowledge about the system analyzed.

Chapter 5- This study is concerned with the use of the Sustainable Development Indicator System (SDIS) to evaluate the performance of Taiwanese distant water fisheries management and related management policies. The four-dimension framework and factor-analysis method are used to aggregate indicators in the SDIS. The ecological index tends towards sustainability in term of exploitation rate and fisheries resources, but the fisheries statistical and biological sampling date are required to assess fish stock more accurately and helpful for set up management measures . The aging of fishing vessels age structure, cost increase and return decrease mean that the economic index tends towards unsustainability. Thus, it is suggested that the Fisheries Agency should formulate policies to increase profits for fishing, improve the operation conditions, and maintain the catch. Due to the rate of foreign fishing crews raised, employment opportunity and social performance have declined, thus the social index tends towards unsustainability. The problem of lack of domestic fishing crews remains. Since the implementation of numerous distant water fisheries management measures in 2006, the institutional index has gradually been moving towards sustainability. Although the management capacity and effectiveness of Taiwanese distant water fisheries has been improving markedly, it is necessary to devote further attention to the problems of management efficiency and cooperation. Lastly, despite the objective of reducing the number of Taiwanese distant water fishing vessels, but the Fisheries Agency has failed to reduce the number of old fishing vessels leading to unsustainability of the fleet due to increasing cost and decreasing profilts.

Chapter 6- Europe 2020: strategy for smart, sustainable and inclusive growth is proposing among other flagship initiatives the resource efficient Europe. Integrated Maritime Policy for the European Union (IMP) considers the Maritime Spatial Planning (MSP) as a fundamental tool for the sustainable development of marine areas and coastal regions, and for the restoration of Europe's seas to environmental health. Governance of marine spaces aims to provide a mechanism for a strategic and integrated plan-based approach for marine management that makes it possible to manage current and potentially conflicting uses. Fishery systems are an obvious example of coupled human and natural systems. A precondition for smart management of the fishery system is the availability of a negative feedback loop based regulatory measure that can constantly work to reduce the difference between the actual and desired states of the system. Fishing feet overcapacity remains the fundamental problem of the Common Fisheries Policy (CFP) while the European fleets remain far too large for the resources available. It is believed that introducing the principles of Europe 2010 Strategy, EU IMP and of the MSP as a key instrument for the IMP implementation, the fishery property rights, e.g. the Individual Transferable Quota (ITQ) system, the Marine Stewardship Council's Principles and Criteria into the Baltic Sea fishery management would contribute to the actual removal of excess fishing capacity. It is suggested that the reversal of burden of proof, when and as necessary, could be technically performed based on the legal procedure of the Environmental Impact Assessment that gives the fishing industry the opportunity to demonstrate that it operates responsibly in return for access to fishing.

Chapter 7- Authors reviewed the marine fisheries laws as contained in the Sea Fisheries Decrees of Nigeria promulgated in 1971, 1972 and revised in 1992. We found out that these decrees were lacking in scientific data needed to effectively enforce the laws, both for industrial and artisanal fisheries. The major habitats of the different species, their breeding and nursery grounds; the migratory movements and seasons for different species; size selection of the fishing gears for different species caught simultaneously by relevant gears; sizes at first maturity of the different species; feeding/trophic relationships of the various species; maximum and optimum sustainable yields of the resource, and the corresponding effort, are some of the missing scientific information. This paper stresses the importance of these scientific data requirements for effective policy options and sustainable fisheries management.

Chapter 8- In 1998, the National Research Council committee on Fish Stock Assessment documented that most fisheries management is based on fishery-dependent forecasting methods. Recent studies have graded the quality of management programs by the amount of fishery-independent data they have on the stock size. We use two applications in Alaska to illustrate contrasting fishery-independent approaches: (1) a direct and precise estimate of the biomass of adult herring, and (2) an indirect study of the conditions in a pink salmon nursery area that could influence survival of the fry to adults. The first, an 18-year study of the herring stock in Prince William Sound, shows that fishery-dependent models twice failed to detect population collapses, and had lag-times of up to five years to detect major population changes. The second, a similar long-term study of the pink salmon stock in Prince William Sound, shows that the adult returns were a function of at least four dynamic biological variables during the spring fry outmigration. In the case of the Pacific herring, we document that the assessment by traditional fishery-dependent models resulted in measures of abundance that were too late, too unrepresentative or too imprecise to make correct management decisions. The failure for timely detection led to overfishing that accelerated a population collapse. The pink salmon example illustrates the case for true ecosystem-based fisheries management. The biological forcing mechanisms that caused change in pink salmon returns were neither understood nor measured. Fisheries-independent data and targeted environmental monitoring offer the best solution to improving the conservation of exploited fish stocks. This approach is a prerequisite for quality fishery management, but must include timeliness, robustness and precision.

Chapter 9- Since the European hake stock is overexploited and presents reduced reproductive capacity, the European Commission (EC) drew up a Recovery Plan, which aims to return this species within biological safety limits in ten years. In this chapter, we will evaluate this management plan, put forward different scenarios and analyse how levels of biomass and catches equilibrium would be affected. A model for considering the size-selectivity of the fishing gears in the dynamic of resource is proposed. The results show that if the less conservationist fleet improves its degree of fishing selectivity, the level for the marine resource will increase and it would possible to place the stock within the limits of safety in ten years (aim of the recovery plan for this stock) and to reach the optimum values. If not, a greater reduction in the TACs should be applied for reaching the limits of biological safety

Chapter 10- The chapter is devoted to dynamic game models related to a bioresource management problem (fishery problems). The main goal of this work is to investigate the schemes to maintain the cooperative exploitation of fish resources. Authors consider different game-theoretic models: discrete-time and continuous, with finite and infinite planning horizon, with different growth rules and players' payoffs functions.

Authors derive the Nash and the cooperative equilibria for considered models. Then authors construct schemes to maintain the cooperation. In the context of cooperative incentive equilibrium we compare traditional method and our approach with changing harvesting territory. In the context of time-consistency we construct imputation set and time-consistent imputation distribution procedure.

In: Fishery Management
Editor: Janice S. Intilli

ISBN: 978-1-61209-682-7

Chapter 1

CORAL REEF FISHERIES AND THE ROLE OF COMMUNITIES IN THEIR MANAGEMENT

Dan A. Exton and David J. Smith
Coral Reef Research Unit
Department of Biological Sciences
University of Essex, Essex, CO4 3SQ, United Kingdom

ABSTRACT

Coral reefs are one of the world's most diverse and productive ecosystems, but are seriously under threat from the problems caused by overfishing. Recent population booms, combined with a low level of economic development make many tropical coastal communities highly dependent on natural resources, leading to extraction rates often significantly exceeding maximum sustainable yield. A lack of suitable alternatives, in terms of both livelihoods and food security, means the use of highly impacting gear types persists, a scenario characteristic of Malthusian overfishing. These economic and socio-biological concerns, typically associated with small-scale artisanal fisheries as seen around coral reefs, bring with them a unique set of problems for conservation managers to overcome. Traditionally in tropical communities, a strict approach of top-down management has been employed, focusing on command and control strategies, but these have tended to fail due, amongst other factors, to a lack of compliance from local stakeholders. Recently, however, there has been increased emphasis on adaptive community-based management, which acts both to empower local user groups, and to build upon the wealth of local ecological knowledge traditional fishing families possess. This chapter aims to introduce coral reef fisheries, describe the distinctive dynamics of coral reef artisanal fisheries, and explore the idea behind community management.

THE IMPORTANCE OF CORAL REEFS AND THEIR FISHERIES

Coral reefs cover around 600,000 km^2 of the world's tropical oceans, 284,300 km^2 of which are found in the shallow coastal waters of over 100 countries (Bryant et al. 1998; Spalding et al. 2001). The majority of these are developing nations, and fisheries associated with coral reefs provide a vital source of food security and other ecosystem services for many tropical coastal communities. Estimates suggest that approximately half a billion people live within 100 km of a coral reef (Whittingham et al. 2003), with varying levels of reliance. The largest group of stakeholders is undoubtedly the millions living subsistence lifestyles, although there are also significant benefits for many people from tourism, and the regional and global trade in reef resources. Added to these are those people who rely on reefs occasionally, as a quasi 'safety net'. One billion people in Asia alone are estimated to be supported by fish catches from coral reefs for at least part of their diets (ICRI 2002). Globally, one billion people receive their major protein source from reef fish (ICRAN 2002), and the livelihoods of half a billion people are directly supported by them (Cesar et al. 1997; Costanza et al. 1997; McManus 1997). This reliance is expected to double by 2050 due to population increases throughout the developing world (McManus 1997).

The total value of coral reefs worldwide has been estimated at US\$ 375 billion yr^{-1} (Costanza et al. 1997), equating to US\$ 6,075 ha^{-1} yr^{-1}, making them one of the most economically valuable ecosystems on the planet. Although quantifying their value in financial terms is difficult (Spurgeon 1992) they act, both directly and indirectly, as an important economic resource in local and global economies (Cesar 2000), with their goods and services currently providing global net revenue of approximately US\$ 30 billion yr^{-1} (Cesar et al. 2003). Obvious benefits are as a source of livelihoods and food security, but reefs also provide important infrastructure services such as coastal protection against erosion and intense storm events such as tsunamis (Berg et al. 1998; Liu et al. 2005). A less obvious benefit, however, is to global biodiversity, with coral reefs one of the most diverse ecosystems in existence (Lewis 1977), home to ca. 25 % of all known marine species. When these benefits are coupled with the other services they provide, including geochemical cycling and potential bioactive compounds for the pharmaceutical industry, one begins to realise that coral reefs are one of the most important ecosystems on the planet (Smith et al. 2007).

The direct value of reef associated fisheries is particularly difficult to calculate, due to their predominantly subsistence nature and a lack of accurate data on landings in many areas, but catch and profit studies have indicated that Southeast Asian fisheries, as an example, produce a net gain of US\$ 2.4 billion yr^{-1}. This is in a region where reliance on reef fish is particularly high, but where many landings go unrecorded due to the nature of subsistence fishing. In the Philippines, they provide 30-40 % of all protein requirements for the population (Bernascek Unpublished) whilst in Indonesia, one of the world's most highly populated countries, over 60 % of the population live within the coastal zone (Elliott et al. 2001), and reef fish provide nearly 70 % of the nation's protein requirements (Cesar et al. 1997; Resosudarmo 2005). The high diversity and productivity of coral reef fish communities are also a major driving factor behind ecotourism, with many tropical nations relying heavily on coral reef tourism for their continued economic development (Smith et al. 2007). As a holistic example, a World Bank study in Montego Bay, Jamaica, investigated the combined economic benefits of coral reefs in an area of only 42.7 hectares. They found that although fishing activity only contributed US\$ 1.31 million, ecotourism from reefs in the area was worth US\$ 315 million, and coastal protection services valued at US\$ 61 million (Ruitenbeek and Cartier 1999; Ruitenbeek et al. 2000), although the latter is difficult to quantify.

A final consideration for financial gains from coral reefs and their fisheries, particularly in future years, is the idea of bioprospecting. Although rather anthropocentric, being based around the precautionary principle that we cannot predict which species will prove beneficial to mankind in the future, bioprospecting describes the search for bioactive compounds with potential in the pharmaceutical market (Cesar 2000; Pendleton 1995; Tamayo et al. 1997).

With the extremely high biodiversity found on and around coral reefs, they represent a potentially major source of success for bioprospecting firms, with the value of reef communities to this industry estimated to be US$ 7775 species^{-1} (Ruitenbeek and Cartier 1999). Political issues from the threat of biopiracy continue to hamper bioprospecting efforts but, if developed carefully, this emerging industry could provide a valuable source of revenue for many developing countries, and is just one more reason why conservation efforts must continue.

ECOSYSTEM THREATS

Despite the numerous services provided by coral reefs and their fisheries, the future of many looks bleak, with anthropogenic and natural stressors set to continue the current trend of reef degradation. Previous studies have suggested that up to 70 % of the world's coral reefs could disappear by 2050 (Cesar et al. 2003), whilst others suggest a complete loss of this vital ecosystem within a century (Smith et al. 2007). A recent assessment by Wilkinson (2008) concluded that 19% of the original area of coral reefs has been lost since 1950, a further 15 % are at risk of being lost in the next 15-20 years and another 20% in 20-40 years. Only 46 % of the world's reefs are healthy and under no immediate threat of destruction (Wilkinson 2008).

Of all the anthropogenic impacts, one of the most direct and hard-hitting is overfishing (McClanahan et al. 2002); a problem which is particularly evident in the Indo-Pacific (Burke et al. 2002; Smith et al. 2007). Coral reefs are known to be highly vulnerable to artisanal fishing pressure (Halpern et al. 2007), and overfishing is commonplace (Coblentz 1997; Exton 2010; Smith et al. 2007). It often provides a catalyst for biodiversity loss (Bellwood et al. 2004; Wilkinson 2008), both through the direct removal of individuals, but also via cascade effects (Bellwood et al. 2004). Coral reef communities are characterised by a complex matrix of ecological interactions, and the loss of a single species can have devastating effects on the ecosystem as a whole, particularly when key functional groups are lost. Algae is kept in check on coral reefs thanks to the oligotrophic nature of tropical marine waters (Larkham 1983), combined with the actions of herbivore populations; a balance severely altered by pollution derived eutrophication (McClanahan et al. 2002; McCook et al. 2001) and by the removal of herbivore populations through overfishing (Mumby et al. 2007). This can lead to algal overgrowth and a subsequent phase shift away from a coral dominated ecosystem (Goreau 1997; Hay and Taylor 1985), greatly reducing the long term carrying capacity of fish populations.

Of particular concern has been the use of destructive fishing techniques. These include blast and cyanide methods, and are generally used to supply the live food and aquarium markets. Cyanide fishing involves the use of potassium cyanide (KCN) or sodium cyanide (NaCN), which mixes with water to form hydrocyanic acid (HCN). When this compound comes into contact with fish at low concentrations, it stuns the individual in a reversible manner (Pet-Soede and Erdmann 2003), allowing fishers to target key valuable species. Unfortunately this technique is responsible for severe reef degradation (Johannes and Riepen 1995; Mous et al. 2000), causing coral bleaching (Jones and Steven 1997), and due to fishers destroying the reef structure in order to collect those individuals seeking refuge within the reef (Pilcher and Oakley 1997). The use of blast fishing involves homemade explosive fertilizer mixtures in glass bottles (Pet-Soede et al. 1999), which are thrown into the water over reefs. The resultant shock wave stuns nearby fish, causing them to float to the surface where they can be collected. Extensive damage to the topographic structure of reefs, an aspect vital to maintaining fish diversity and productivity, is caused by the blast, leaving only coral rubble (Edinger et al. 1998; McManus 1997). Even five years after blast damage, no significant recovery of the reef is typically seen (Fox et al. 2003). Previous estimates of destructive fishing damage to reefs have estimated that, in Indonesia alone, 0.06 % of total reef area is destroyed annually by cyanide fishing (Mous et al. 2000) and 3.75 % annually by blast fishing (Pet-Soede et al. 1999), although more recent reports suggest that the problem is fading in some areas (Exton 2010; Pet-Soede and Erdmann 2003).

Reduction in fishing pressure, both destructive and non-destructive, forms a major priority for conservation organisations around the world (Wilkinson 2008). Direct anthropogenic impacts are driving severe declines in reef diversity globally (Bellwood et al. 2004; Mora et al. 2006), while the ability of species to cope with natural disturbances is known to be greatly reduced by continued exploitation (Chape et al. 2005; Palumbi 2004; Sale et al. 2005). Added to this is the complex nature of habitat connectivity, meaning that negative impacts affecting linked systems have knock-on effects for coral reefs, particularly their fisheries. Many reef associated fish species do not exist exclusively on coral reefs, but in fact spend significant periods of their life cycles in other tropical coastal habitats, predominantly mangroves and seagrass beds (Nagelkerken et al. 2002; Nagelkerken et al. 2000; Unsworth and Cullen 2010; Unsworth et al. 2007a). Studies have highlighted the importance of the presence of these connected habitats in determining the community assemblage found on reefs themselves (Mumby 2006), and means that both mangrove and seagrass habitats are vital to the success of coral reef fisheries (Unsworth et al. 2007b). Unfortunately, the loss of these important ecosystems is also taking place, through anthropogenic factors including coastal development and the exploitation of mangrove for firewood and building materials (Unsworth and Cullen 2010; Unsworth et al. 2007a; Unsworth et al. 2007b; Waycott et al. 2009; Wilkinson 2008), disrupting the mangrove-seagrass-reef continuum and impacting on nearby fisheries.

Fishery Dynamics

The high productivity of reef fisheries supports mostly subsistence lifestyles, and commercial fishing activity is much less common. Reef fisheries are therefore characterised by mostly artisanal techniques, many of which involve non-selective gear types (Exton 2010). In total, artisanal activity on reefs accounts for 50 % of fish consumed by humans worldwide (Matthew 2001). These artisanal fisheries tend to be multi-gear, including the use of nets, traps, line, and spear gunning (Acosta 1994; Davies et al. 2009; Exton 2010; McClanahan and Mangi 2004; Pet-Soede and Erdmann 2003; Weis and Weis 2005). In some regions, fishery aggregation devices (FADs) are also employed whereby temporary structures, moored in deeper water, provide a physical complexity which attracts reef fish populations, an activity promoted by many conservation organisations as a sustainable alternative to direct reef fish extraction. Invertebrate communities are also commonly exploited, typically through intertidal gleaning activity on seagrass beds and reef flats (Ashworth et al. 2004; Dalzell et al. 1996; De Boer and Prins 2002; Gladstone 2000; Munro 1989). Although certain species are prized for their financial value at market, there is little concept of bycatch amongst subsistence fishers, and juvenile catches are often high (Exton 2010).

As well as being multi-gear, reef fisheries also tend to utilise a large number of landing sites (Exton 2010). Compared to the more centralised nature of temperate commercial fisheries, those incorporating small tropical coastal communities tend to utilise any coastal settlement as a potential landing site. This poses significant problems for the policing of these fisheries, as accurate monitoring of landings is close to impossible in many regions. Similarly, catches are not distributed in a simple market environment. Informal markets are commonplace, and in many cases the majority of a catch will be shared between family and friends before any remainder is sold to the wider community (Exton 2010). Although some commercial activity takes place, this is often seasonal and targets specific catches (Pet-Soede and Erdmann 2003).

With a few exceptions, the majority of nations with coral reefs within their borders are considered developing countries. Therefore inherent problems commonplace in the developing world play an enormous role in the state of natural ecosystems and levels of exploitation. These social, economic and political factors are considered to be amongst the most important issues contributing to the overall threat to natural systems globally (Smith et al. 2007). One of these is poverty, a key driver behind the high reliance on natural resources for food security. The direct relationship between poverty and a deterioration of the environment has been long understood (Brundtland 1987), and is clearly seen with regards to coral reef fisheries in the form of Malthusian overfishing (Pauly 1994).

Although overfishing is case specific, coral reefs are particularly susceptible to its Malthusian form, which is defined as "the situation caused by poor fishers when, faced with declining catches, induce wholesale resource destruction in an attempt to maintain income" (McClanahan et al. 2008; Pauly 1988; Pauly 1990; Pauly 1994; Pauly et al. 1989). The results of this process can take a number of forms, ranging from the most extreme, where destructive extraction techniques such as dynamite and cyanide fishing are employed, to simply unsustainable exploitation including significant avoidable juvenile catches and severe stock depletion with negative consequences for ecosystem health and function. The common driving forces behind Malthusian overfishing include population growth leading to significant increases in the number of fishers in an area, and a lack of economic and food security alternatives available to local communities (McManus 1997; Pauly 1994). In recent years there is also a problem caused by the aspirations of artisanal fishers for improved living conditions, which acts as a disincentive to comply with conservation (Randall 1991).

The persistence of poverty is a particular problem in subsistence-based fisheries, as it seriously impacts the ability of resource users to exit a fishery in a state of decline (Adato et al. 2006; Carter and Barrett 2006; Cinner et al. 2008). These so-called 'poverty traps' are driven by an inability amongst poorer fishing families to enter higher-return income streams (typically more professional occupations) and are therefore forced to remain in the fishery even through stock declines and drops in catch quality and size (Cinner et al. 2008; Dasgupta 1997). Consequently, subsistence fisheries in developing countries have experienced particular difficulty in achieving sustainability and successful fishery and ecosystem management (Cacaud 2004; Grafton et al. 2008).

The role of Marine Protected Areas

In light of the extensive threats to the health and sustainability of coral reef fisheries, management of these important natural resources is of key importance. It is widely understood that a move to sustainable fisheries will require a number of changes in the way extraction takes place, including a significant reduction in fishing effort, addressing the problem of bycatch, and an end to fishing techniques which cause widespread harm to the ecosystem as a whole (Hilborn 2007b). An investigation by the Food and Agriculture Organization (FAO) concluded that, for this to be achieved, emphasis needs to be placed on good governance, appropriate incentives, a reduction in demand for catches, and poverty alleviation, amongst others (Greboval 2002; Greboval 2004).

But the case of small-scale artisanal fisheries is a difficult one to manage, as there is such a conflict between the classic categories of management objectives: biological, economic, social and political (Hilborn 2007b). To satisfy any one of these categories requires another to be overlooked, with serious consequences for either ecological health or the wellbeing of local communities, leaving management bodies with a difficult dilemma. To satisfy biological criteria, for example, extraction rates should ideally be reduced to zero, whereas social and political objectives require increased employment levels, and therefore demand maximised fishing effort. A ‘happy medium’ is therefore of upmost importance if management is to be successful. If this is achieved, however, the potential benefits are vast, and include protection for biodiversity and ecosystem health, improved food security, and greater economic and social benefits to fishing communities (Grafton et al. 2008).

One of the most widely employed management techniques to promote fishery sustainability is the use of marine protected areas (MPAs), which are widely regarded as one of the simplest ways to manage overexploited fisheries (Roberts 1997; Roberts and Polunin 1993). This approach has formed the basis of orthodox management strategies targeting reef fisheries, and involves top-down policies largely based on stock assessments (Allison and Ellis 2001). By protecting fish stocks from overexploitation, a number of benefits are provided to the fishery. These are primarily in terms of stock recovery, which has significant ecological benefits including increases in biomass, abundance, diversity and average size (Roberts et al. 2001), and improved resilience provided by increases in the populations of key functional groups (Bellwood et al. 2004; Mora et al. 2006; Mumby and Harborne 2010). This stock increase within MPAs also provides benefits to adjacent fishing grounds through a ‘spill-over’ of adults, larvae and eggs from recovered populations (Harmelin-Vivien et al. 2008). They also provide an important source for the export of pelagic larvae and eggs (Gell and Roberts 2003). If extraction rates on adjacent reefs within this buffer zone are then maintained at or below the maximum sustainable yield (MSY), catch per unit effort (CPUE) and profits to fishers can be greatly enhanced (Guidetti and Claudet 2010; Pauly et al. 2002). Multi-zoned management strategies utilising a combination of strategically placed no-take areas with carefully managed fishing grounds can therefore provide an area with the means of achieving fisheries sustainability, although complex zonation plans may struggle to gain community support and compliance.

Despite the potential benefits, the effectiveness of MPAs in poor tropical countries has been questioned (McClanahan 1999). Previous studies have shown that of 1300 MPAs worldwide incorporating intertidal components (Kelleher et al. 1995), 71 % have “unknown management" status (McClanahan 1999). Therefore the majority can be termed merely 'paper parks', where cooperation is not rewarded, nor violation punished (Rudd et al. 2003); thus failing to manage resources with any degree of success (Alcorn 1993). This is typically due to a lack of good management and enforcement following MPA establishment (Burke and Maidens 2004; Burke et al. 2002). In 1998, 46 MPAs were designated in the Western Indian Ocean, excluding the Red and Arabian Seas (Salm and Ngoile 1998). Of these only eight, totaling a mere 40 km^2, were reported to successfully restrict the extraction of reef resources. Those around Dar es Salaam, for example, had between 2 - 5 fishers km^{-2} (McClanahan et al. 1999b), whilst almost all of the 15 MPAs around the Seychelles had between 3.5 – 20 fishing boats km^{-2} (Jennings et al. 1996). Tanzania shows similarly low levels of success (Muthinga et al. 2000), and in total below 10% of Western Indian Ocean MPAs had achieved successful results by the turn of the century (McClanahan 1999). Globally, 18.7 % of coral reefs (98,650 km^2) are within a network of 980 MPAs, but less than 0.1 % of the world's reefs are within MPAs where a no-take policy is strictly enforced (Mora et al. 2006).

In spite of these concerns, MPAs remain an important component of reef fishery management. Once they have made it through the initial implementation stages, with adequate management efforts they can be both ecologically and economically successful (McClanahan 1999). The network of three Kenyan MPAs are testament to this, with over 25 years of full protection in which time substantial income has been raised from visitors. Research indicates that fish populations and reef health have been maintained, whilst surrounding reefs not under protection show extensive coral rubble and algal overgrowth (McClanahan and Obura 1995; McClanahan and Shafir 1990). Further positive effects on commercially important fish and invertebrate species, as well as aspects of reef ecology including an improved calcium carbonate budget, have been reported (McClanahan 1995). Despite these benefits, however, there are also potential concerns in Kenyan MPAs, and future proliferation of MPAs throughout the tropics will need to be carefully considered. The revenue from visitors to protected Kenyan reefs has begun to decline after the initial boom in visitor numbers, whilst environmental factors from beyond the parks' boundaries threaten their future health. The Malindi MPA is regularly covered in silt (McClanahan and Obura 1997), whilst the Watamu MPA receives a high nutrient load (McClanahan et al. 1999a), both as a result of land use changes. These external factors are amongst the reasons why many advocates of MPA use also support the idea of integrated coastal zone management (ICZM); a holistic approach to reef management whereby protection incorporates terrestrial coastal areas as well as linked marine habitats along the reef-seagrass-mangrove continuum (Christie et al. 2005; Unsworth and Cullen 2010).

Unfortunately, if MPAs are to make a significant contribution to the future sustainability of coral reefs and their fisheries, their coverage and performance must both be significantly improved (Mascia 2003). Although there have been calls for an increased use of MPAs in management (Beddington et al. 2007), they can only form one part of a much broader management effort. Their designation must be followed by a more specific management strategy, often tailor-made to address the particular challenges of an area. With the protection of food security, livelihoods and cultural lifestyles a key component of management aims, and the importance of compliance amongst stakeholders if success is to be achieved, any management attempts should ultimately benefit the fishing community as well as the ecosystems themselves (Dalton 2010).

Adaptive Community Management

Although it is generally accepted that the use of MPAs provides the foundation for successful fisheries management (Beddington et al. 2007), many experts also agree that, if used in isolation, all MPAs will achieve is a displacement of fishing effort elsewhere (Grafton et al. 2006; Halpern et al. 2004; Roberts et al. 2005; Stefansson and Rosenberg 2005). Also likely to take place, in the absence of strict enforcement, is continued fishing effort in the form of poaching, as local communities struggle to maintain food security in the face of no-take policies. These issues are often overlooked, as many MPA plans make the assumption that fishing effort will simply disappear (Wilen et al. 2002). In reality, the primary determinants of MPA success have been shown to be social factors (Kelleher and Recchia 1998; McClanahan 1999), and therefore the most effective MPAs are those with a collaborative approach to management (Mascia 2003). Resource extraction is human-led, and management must approach it as such (Fox et al. 2006; Richardson et al. 2006), with food security and livelihoods given adequate consideration.

Adaptive community management (or co-management) allows fishing effort to be controlled on a finer scale, through bottom-up methods giving emphasis to the social component of natural resource management. It aims to shift management authority away from the conventional groups of government bodies and non-governmental organisations (NGOs), and towards local communities (Fabricius and Collins 2007). This can fit into the broader structure of MPA use, whereby a devolution of decision-making powers away from national and regional government can boost MPA performance through the empowerment of local resource users and other stakeholders (Fabricius and Collins 2007; White et al. 2002). Essentially, co-management provides an approach which aims to link natural resources with the interests of local communities (Mascia 2003), thus addressing key concerns attributed to the use of stricter top-down management strategies alone, primarily compliance (Claudet and Guidetti 2010; Pretty and Smith 2004) and effort displacement (Grafton et al. 2006; Grafton et al. 2008; Halpern et al. 2004; Hilborn 2007a; Hilborn 2007b).

According to Grafton et al. 2006, there are four main tools which can be built into a co-management approach to increase the likelihood of success: rights, participation, capacity building and benefit distribution. The specific aspects of these will vary from case to case (Hilborn 2007b), but they provide a framework for conservation managers. The fundamental basis of all four categories, however, is to provide appropriate incentives to resource users to encourage a sustainable approach to exploitation (Hilborn 2007a; Hilborn 2007b).

The traditionally open access nature of coral reef subsistence fisheries, often termed 'Olympic' systems, induces a race to fish mentality amongst fishing communities which drives increases in effort, often regardless of ecosystem consequences. Moving away from an Olympic fishery is therefore a primary aim of managers, which can be achieved by the introduction of dedicated access (Hilborn 2007b). Particularly in developing countries, where national infrastructure is often lacking in terms of conservation support, well-defined secure access rights for local fishermen is a key component in the early management phase (Grafton et al. 2008; Hilborn 2007b). These measures remove any existing pressure from external fishers, and encourage a sense of ownership amongst local stakeholders, thus promoting the idea of sustainability and conservation. Importantly, community empowerment in the regional economy is also significantly boosted (Pomeroy et al. 2001).

In some situations, the idea of restricted access for reef fisheries has been taken a step further through the leasing of submerged lands. In Raja Ampat, Indonesia, for example, communities legally own the coastal waters nearby, and lease them to fishers and pearl harvesters (Beck et al. 2004). In another example, certain villages in the Wakatobi Marine National Park, Indonesia, have a lease agreement with a local dive tourism resort, whereby communities receive payment not to engage in resource extraction on certain reefs (Beck et al. 2004). An added benefit of engendering a sense of ownership for reef resources amongst local user groups is that it encourages self-policing, whereby 'foreign' fishing effort is identified and dealt with by the community, without the need for expensive enforcement programs at government or NGO level. In many managed fisheries, the ability to identify and punish illegal fishing effort is a major challenge (Beddington et al. 2007), and therefore the possibility of self-policing is an attractive one.

Participatory management techniques form the backbone of co-management approaches to reef fisheries, and apply to all aspects of the management effort. As well as sharing decision making authority (Fabricius and Collins 2007; Grafton et al. 2006), stakeholders should also be included in research and monitoring aspects (Mascia 2003). It is well recognised that resources can be better managed when stakeholders participate in the management process (Pomeroy 1995). For truly participatory management to be possible, however, government decentralisation is critical, as it localises legislative power and encourages tailor-made management strategies (Pomeroy and Berkes 1997). In the Philippines, for example, formal decentralisation occurred thanks to the Local Government Code (1991), which transferred responsibilities to local government and communities (Pomeroy et al. 2001). Equally important however is for governance to be effective, and appear transparent to stakeholders (Hilborn 2007b). With good governance in place, participatory co-management typically involves a 'trialogue' of parties: local communities, scientists/NGOs, and government (Fabricius and Collins 2007). Each group bring their own strengths to the management cooperative, and play an essential and, importantly, meaningful role (Grafton et al. 2006; Guidetti and Claudet 2010). Scientists and NGOs often have a particularly important role to play as change agents, acting independently as quasi consultants to advise and train local communities where required (Pomeroy et al. 2001).

One important aspect of the participatory nature of co-management is the opportunity for local ecological knowledge to be incorporated into the decision making process. Generations spent in subsistence lifestyles have led local fishers to develop a keen understanding of the environmental impact of their exploitation (Drew 2005), and in many cases have subsequently developed their own successful self-management practices (Allison and Ellis 2001; Feit 1988). Many of these are incorporated into local culture and beliefs (Rudd et al. 2003), as per the practice in parts of Papua New Guinea, where access to certain areas of reef are restricted during mourning periods, often lasting many years (Wright 1985), and thus act as a period of ecosystem regeneration and stock recovery. Many command and control management strategies tend to exclude reef stakeholders and undermine traditional management practices (Smith et al. 2007), but co-management provides a method for their inclusion. Another key benefit of stakeholder participation is the development of social capital, which refers to aspects of community structure such as networks and trust which encourage and facilitate cooperation to achieve common goals (Cundill and Fabricius 2010). Providing an open relationship between local fishers, government and conservation organisations, where their opinions are not only heard but also incorporated into management, is an ideal way to build trust. This helps ease the scepticism felt towards scientists and conservationists by many reef fishing communities (Guidetti and Claudet 2010), and greatly increases levels of compliance (Pretty and Smith 2004), which is essential for fisheries success (Claudet and Guidetti 2010; Hilborn 2007a).

Our initial thoughts on capacity building are often of education programs for local stakeholders, and this should certainly form a critical part of the process. Improving local understanding and awareness of environmental issues, particularly sustainability, can only improve compliance with management, whilst feedback on the successes of management to user groups will greatly enhance performance and success (Gelcich et al. 2008; Kritzer 2004; Pollnac et al. 2001a). The empowerment of fishing communities is also further enhanced through programs which increase their knowledge base (Pomeroy et al. 2001), building on the benefits provided by restricted access. But capacity building should also benefit other members of the management trialogue, particularly scientists and NGOs. In fact, a fundamental component of reef fishery management should be the collection of baseline data through well-funded programs focusing on the resource being exploited (Hilborn 2007b; Mascia 2003). This provides performance monitoring data, which can be used to identify areas of success and failure within the objectives of the management process, and facilitate an adaptive strategy. Data collection can take many forms, but will often include regular monitoring of landings to assess changes in catch statistics, as well as environmental monitoring to identify changes in ecosystem health. In addition, original research programs can greatly enhance the knowledge on which management strategies are based. These can take the form of biological research into the complex nature of coral reef ecosystems (Hilborn 2007b), or socioeconomic studies focusing on the dynamics of fishing livelihoods (Cinner et al. 2008). To maintain the cooperative nature of co-management however, both monitoring and research programs should be participatory wherever possible (Mascia 2003).

The success of co-management is dependent on the premise that the benefits offered by management, as well as the costs, are shared equally amongst all stakeholders (Pomeroy et al. 2001). If individuals or groups feel they are being unfairly treated by the management strategy, or that compensation is inadequate, positive changes in the form of reduced catches can be impeded (Grafton et al. 2008). It is important for other members of the management trialogue to understand the risk involved in exiting the fishery for families reliant on a subsistent lifestyle, and measures to compensate losers must minimise this risk. The cooperative nature of co-management should go a long way towards addressing these concerns; in so much as it should encourage an understanding of the management strategy amongst fishers, as well as gaining their support. This approach is not typically enough, however, and compensation mechanisms must be in place from the beginning to encourage resource users to exit the fishery, and support them thereafter (Grafton et al. 2008). Economic incentives in the form of tradable licenses can provide families with the capital needed to invest in alternative livelihoods, but key to successfully reducing fishery exploitation in poor tropical coastal communities is the ability to provide the opportunity for income diversification, a topic discussed in greater detail shortly.

The community-based management of fisheries has been in place in various locations around the world for the past few decades, with varying levels of success (Pomeroy et al. 2001). One successful program in the Mediterranean has reported significant increases in CPUE and the presence of valuable species after only three years of community-managed exploitation (Guidetti and Claudet 2010), when compared to declines in areas without co-management in place (Alcala et al. 2005; Ferraris et al. 2005). In tropical regions, a number of co-management strategies are in place to promote sustainability in coral reef fisheries, showing promising results. In Samoa, 44 villages have produced management plans to date, with 38 establishing self-owned and managed MPAs (King and Faasili 1999). In fact, community-based management of reef fisheries is growing throughout many Pacific islands, incorporating a range of methods including restricted access, closed seasons, and gear restrictions (Johannes 2002). In Hawaii, there has been significant progress towards co-management in combination with MPA designation (Tissot et al. 2009), whilst there has been growing community involvement in fisheries management planning in Thailand, Vietnam, Cambodia and Malaysia (Nasuchon and Charles 2010). Comparatively, the move from successfully run co-management to centralized state-run management in the 1990s around Apo Island, Philippines, previously described as one of the best examples of co-management of marine resources anywhere in the world, has since lost the support of stakeholders (Hind et al. 2010).

But despite these promising signs, co-management practices remain fraught with difficulties. As such a dynamic and site-specific process, the opportunities for error and failure are significant, and can come in many forms. Many of these potentially negative factors acting on co-management are summarised by Fabricius and Collins (2007), but include financial mismanagement (Robbins 2000), unfavourable change at the political level (Biggs et al. 2004), and an abuse of restricted access rights by stakeholders (Okello and Kiringe 2004). Others are difficult to avoid, such as a loss of revenue from tourism due to changes in the global market (Fabricius and Collins 2007), but can induce an increase in fishing effort as employment in non-resource based sectors decreases. Providing the co-management process remains adaptive, however, obstacles can be overcome and the future sustainability of coral reef subsistence fisheries achieved.

ALTERNATIVE LIVELIHOODS

Alternative income streams for resource users are an integral component of management attempts for coral reef fisheries. Primarily, they are vital in preventing the displacement of fishing effort which is common under marine resource management (Halpern et al. 2004), whilst aiding in the reduction of fishing effort within management borders. A common consequence of a reduction in fishing effort is reduced employment, and the livelihoods of those fishers exiting the fishery must be taken care of. Without this provision, there will be a lack of sufficient incentives for stakeholders to surrender their right to fish, and management will simply not succeed. A one-off financial compensation payment, as received under a tradable licensing scheme, may temp some fishers to suspend their fishing activities, but without a means of providing for themselves and their families in the long term, this will only prove temporary.

Under normal conditions, artisanal fishing is one of the few livelihoods available to poor coastal communities in the developing world, forming a major driver of Malthusian overfishing in these regions (Pauly 1994). Providing alternatives to fishing therefore also helps eliminate widespread poverty and facilitates economic development, by assisting poorer families to escape the poverty trap many find themselves in (Costanza 1987). Importantly, however, is not the simple provision of alternative options, but that the occupations available satisfy the non-economic aspects of a fishing livelihood (Pollnac and Poggie 2006; Pollnac et al. 2001b). A fishing lifestyle often forms an important cultural component of the lives of fishing communities, for example time spent on the water, and alternative livelihoods should satisfy these aspects wherever possible if they are to be incentive enough to leave the fishery.

Previous work has found that many fishers in low-income countries already demonstrate a high diversity of livelihoods (Allison and Ellis 2001). In some cases, therefore, fishers will already pursue non-resource based occupations as a means of supporting themselves during seasonal periods of reduced catch or accessibility. In most cases, however, conservation and government bodies (i.e. the non-community members of the management trialogue) will be required to develop alternative income streams which fit in with the management objectives. In these cases, fishing communities need to be provided with the equipment and expertise required to generate a healthy income, often by providing families with products of commercial value. Ideally, this should be incorporated into the initial stages of management efforts, as developing ideas into realistic working options can take considerable time.

Probably the most common example of an alternative income stream is mariculture: the production of marine organisms under controlled conditions, which are harvested and sold for a profit. Although significant financial gain can be made from 'closed mariculture', an expensive and technologically advanced technique, subsistence communities are typically better suited to 'open mariculture', which is relatively simple and inexpensive to implement and maintain. A variety of organisms are used in mariculture by reef-dependent communities, including invertebrates such as Tridacnidae (Lucas 1994) and algae such as the carrageenophytes (Schiel and Nelson 2004), as well as fish and shrimp species.(Suvapepun 1994).

The farming of seaweed is a common feature around reefs, particularly in Southeast Asia, and is considered to be a useful alternative livelihood to fishing (Sievanen et al. 2005). There is large scale global demand for algae, predominantly the thickening qualities of derived hydrocolloids, with some estimating its value at ca. US$ 800 million year^{-1}. At present, 99 % of carrageenophytes produced for the global market are cultivated in the tropics, particularly in the Philippines, Indonesia, Malaysia and Tanzania (Blankenhorn 2007). It is a favoured alternative income stream for management bodies due to its minimal environmental impacts, thanks to the absence of fertilisers and pesticides (FAO 2010). One concern regarding its suitability as an incentive in fishery management is the uneven distribution of profits within the industry. Seaweed farmers often enter the industry via loans from middlemen, who also purchase the raw product from the farmers for a relatively low price. The high profits are only obtained at the processing stage, during hydrocolloid production at processing plants. Management should therefore consider subsidising the industry, perhaps via a compensation package for surrendering fishing licenses. Another possibility is the formation of local cooperatives able to enter the hydrocolloid market itself, bypassing the middlemen and thus maximising profits for themselves. This would also boost the income of those individuals already involved in seaweed farming, via increased prices for existing goods.

Another commonly employed mariculture technique in tropical coastal communities is shrimp farming. The major benefits are its economic viability: in 1995 Ecuador exported US$ 668 million worth of shrimp, a level of production repeated in many countries worldwide (Suvapepun 1994). Unfortunately, however, there are significant ecological impacts associated with shrimp farming. Wastewater, fertilised with urea and triple superphosphate during the farming process, is often discharged directly into the coastal marine environment, with serious consequences for local ecosystems, particularly coral reefs. In many locations, farms also require mangrove clearance to make room for mariculture sites, damaging the mangrove-seagrass-reef continuum (Unsworth et al. 2008). Another viable alternative for families exiting the fishery is the ecotourism industry. Revenues range from the provision of SCUBA diving and snorkelling activities, to accommodation and food, and can form a major contributor to many small tropical economies (Cesar et al. 2002; Patterson et al. 2004; White et al. 1997). One of the key difficulties with ecotourism as an alternative to resource extraction is the difficulty in maintaining profits within the local economy. Additional concerns include potential fluctuations in the tourism industry due to external factors in the global economy (Fabricius and Collins 2007). If either of these concerns come to pass, the most likely outcome is a return to fishing for those families who have lost out.

Conclusion

The productive fisheries associated with coral reef ecosystems face a host of threats to their survival. Principal of these is extreme overfishing, particularly the Malthusian form driven by socioeconomic factors commonplace in subsistence communities. A heavy reliance on natural resources, combined with increasing human populations and a lack of alternatives, means pressure on reef fisheries is set to continue. However, with suitable management strategies in place, extraction rates can be maintained below maximum sustainable yield, whilst social and economic development assists local stakeholders to progress from ecologically damaging subsistence lifestyles. Strict command and control strategies involving no-take MPA designation remains in use around the world, but recent decades have seen a shift towards adaptive community-based management. MPAs remain an important management tool, providing the basis of partial protection, as well as a framework at government and NGO level to pursue co-management. Added to this top-down strategy is the provision of incentives to encourage resource users to exit the fishery, including restricted access engendering a sense of ownership, as well as community education programs and the development of alternative income streams. Specific management plans will vary on a site by site basis, and progress can take time, but providing the process remains adaptive, reacting to problems and failings which are encountered along the way, long term success is achievable.

References

Acosta, A. R. 1994. Soak time and net length effects on catch rate of entangling nets in coral reef areas. *Fisheries Research* 19: 105-119.

Adato, M., M. R. Carter, and J. May. 2006. Exploring poverty traps and social exclusion in South Africa using qualitative and quantitative data. *Journal of Development Studies* 42: 226-247.

Alcala, A. C., G. R. Russ, A. P. Maypa, and H. P. Calumpong. 2005. A long-term, spatially replicated experimental test of the effect of marine reserves on local fish yields. Canadian *Journal of Fisheries and Aquatic Sciences* 62: 98-108.

Alcorn, J. B. 1993. Indigenous peoples and conservation. *Conservation Biology* 7: 424-426.

Allison, E. H., and F. Ellis. 2001. The livelihoods approach and management of small-scale fisheries. *Marine Policy* 25: 377-388.

Ashworth, J. S., R. F. G. Ormond, and H. T. Sturrock. 2004. Effects of reef-top gathering and fishing on invertebrate abundance across take and no-take zones. *Journal of Experimental Marine Biology and Ecology* 303: 221-242.

Beck, M. W., T. D. Marsh, S. E. Reisewitz, and M. L. Bortman. 2004. New tools for marine conservation: the leasing and ownership of submerged lands. *Conservation Biology* 18: 1214-1223.

Beddington, J. R., D. J. Agnew, and C. W. Clark. 2007. Current problems in the management of marine fisheries. *Science* 316: 1713-1716.

Bellwood, D. R., T. P. Hughes, C. Folke, and M. Nyström. 2004. Confronting the coral reef crisis. *Nature* 429: 827-833.

Berg, H., M. C. Öhman, S. Troëng, and O. Lindén. 1998. Environmental economics of coral reef destruction in Sri Lanka. *Ambio* 27: 627-634.

Bernascek. Unpublished. *The role of fisheries in food security in the Philippines: a perspective study for the fisheries sector to the year 2010.*

Biggs, R. and others 2004. Nature supporting people The Southern African Millennium Ecosystem Assessment Integrated Report. *Council for Scientific and Industrial Research.*

Blankenhorn, S. U. 2007. *Seaweed farming and artisanal fisheries in an Indonesian seagrass bed: Complementary or competitive usages?* University of Bremen.

Brundtland, H. 1987. *Our common future.* Oxford University Press.

Bryant, D., L. Burke, J. W. McManus, and M. D. Spalding. 1998. *Reefs at risk: a map-based indicator of threats to the world's coral reefs.* World Resources Institute.

Burke, L., and J. Maidens. 2004. Reefs at risk in the Caribbean. World Resources Institute.

Burke, L., E. Selig, and M. Spalding. 2002. *Reefs at Risk in Southeast Asia.* World Resources Institute.

Cacaud, P. 2004. The case of developing countries. *FAO Fisheries report.* FAO.

Carter, M. R., and C. B. Barrett. 2006. The economics of poverty traps and persistent poverty: an asset-based approach. *Journal of Development Studies* 42: 178-199.

Cesar, H. S. J. 2000. Coral reefs: their function, threats and economic value, p. 14-39. *In* H. S. J. Cesar [ed.], Collected essays on the economics of coral reefs. *CORDIO.*

Cesar, H. S. J., L. Burke, and L. Pet-Soede. 2003. The Economics of Worldwide Coral Reef Degradation. *Cesar Environmental Economics Consulting Report* 23.

Cesar, H. S. J., C. G. Lundin, S. Bettancourt, and J. Dixon. 1997. Indonesian coral reefs: an economic analysis of a precious but threatened resource. Ambio 26: 345-350.

Cesar, H. S. J., P. Van Beukering, S. Pintz, and J. Dierking. 2002. Economic valuation of the coral reefs of Hawaii. *Hawaii Coral Reef Initiative Research Program.* University of Hawaii.

Chape, S., J. Harrison, M. Spalding, and I. Lysenko. 2005. Measuring the extent and effectiveness of protected areas as an indicator for meeting global biodiversity targets. *Philosophical Transcriptions of the Royal Society B* 360: 443-455.

Christie, P. and others 2005. Key Findings from a Multidisciplinary Examination of Integrated Coastal Management Process Sustainability. *Ocean and Coastal Management* 48: 468-483.

Cinner, J. E., T. Daw, and T. R. McClanahan. 2008. Socioeconomic Factors that Affect Artisanal Fishers' Readiness to Exit a Declining Fishery. *Conservation Biology* 23: 124-130.

Claudet, J., and P. Guidetti. 2010. Fishermen contribute to protection of marine reserves. *Nature* 464: 674.

Coblentz, B. E. 1997. Subsistence consumption of coral reef fish suggests non-sustainable extraction. *Conservation Biology* 11: 559-561.

Costanza, R. 1987. Social traps and environmental policy. *BioScience* 37: 407-412.

Costanza, R. and others 1997. The value of the world's ecosystem services and natural capital. *Nature* 387: 253-260.

Cundill, G., and C. Fabricius. 2010. Monitoring the Governance Dimension of Natural Resource Co-management. *Ecology and Society* 15.

Dalton, R. 2010. Reserves 'win-win' for fish and fishermen. *Nature* 463: 1007.

Dalzell, P., T. J. H. Adams, and N. V. C. Polunin. 1996. Coastal fisheries in the South Pacific Islands. *Oceanography Marine Biology Annual Review* 34: 395-531.

Dasgupta, P. 1997. Nutritional status, the capacity for work, and poverty traps. *Journal for Econometrics* 77: 5-37.

Davies, T. E., N. Beanjara, and T. Tregenza. 2009. A socio-economic perspective on gear-based management in an artisanal fishery in south-west Madagascar. *Fisheries Management and Ecology* 16: 279-289.

De Boer, W. F., and H. H. T. Prins. 2002. Human exploitation and benthic community structure on a tropical intertidal mudflat. *Journal of Sea Research 4*8: 225-240.

Drew, J. A. 2005. Use of traditional ecological knowledge in marine conservation. *Conservation Biology* 19: 1286-1293.

Edinger, E. N., J. Jompa, G. V. Limmon, W. Widjatmoko, and M. J. Risk. 1998. Reef degradation and coral biodiversity in Indonesia: effects of land based pollution, destructive fishing practices and changes over time. *Marine Pollution Bulletin* 36: 617-630.

Elliott, G., B. Mitchell, B. Wiltshire, I. A. Manan, and S. Wismer. 2001. Community participation in marine protected area management: Wakatobi National Park, Sulawesi, Indonesia. *Coastal Management* 29: 295-316.

Exton, D. A. 2010. Nearshore fisheries of the Wakatobi, p. 193-207. In J. Clifton, R. K. F. Unsworth and D. J. Smith [eds.], *Marine research and conservation in the coral triangle:* the Wakatobi National Park. NOVA

Fabricius, C., and S. Collins. 2007. Community-based natural resource management: governing the commons. *Water Policy* 9: 83-97.

FAO. 2010. Cultured aquatic species information programme - *Eucheuma spp.* FAO Fisheries and Aquaculture Department [online].

Feit, H. A. 1988. Self-Management and State-Management: Forms of Knowing and Managing Northern Wildlife, p. 79-91. In M. M. R. Freeman and L. N. Carbyn [eds.], *Traditional Knowledge and Renewable Resource Management in Northern Regions.* Boreal Institute for Northern Studies, and International Union for the Conservation of Nature and Natural Resources.

Ferraris, J., D. Pelletier, M. Kulbicki, and C. Chauvet. 2005. Assessing the impact of removing status on the Abore Reef fish assemblage in New Caledonia. *Marine Ecology-Progress Series* 292: 271-286.

Fox, H. E., C. Christian, J. C. Nordby, O. R. W. Pergams, G. D. Peterson, and C. R. Pyke. 2006. Perceived barriers to integrating social science and conservation. *Conservation Biology* 20: 1817-1820.

Fox, H. E., J. S. Pet, R. Dahuri, and R. L. Caldwell. 2003. Recovery in rubble fields: long-term impacts of blast fishing. *Marine Pollution Bulletin* 46: 1024-1031.

Gelcich, S., M. J. Kaiser, J. C. Castilla, and G. Edwards-Jones. 2008. Engagement in comanagement of marine benthic resources influences environmental perceptions of artisanal fishers. *Environmental Conservation* 35: 36-45.

Gell, F. R., and C. M. Roberts. 2003. Benefits beyond boundaries: the fishery effects of marine reserves. *Trends in Ecology and Evolution* 18: 448-455.

Gladstone, W. 2000. The ecological and social basis for management of a Red Sea marine-protected area. *Ocean & Coastal Management* 43: 1015-1032.

Goreau, T. J. 1997. Bleaching and reef community change in Jamaica: symposium on long term dynamics of coral reefs. *American Zoologist* 32: 683-695.

Grafton, R. Q. and others 2006. Incentive-based approaches to sustainable fisheries. Canadian *Journal of Fisheries and Aquatic Sciences* 63: 699-710.

---. 2008. Positioning fisheries in a changing world. *Marine Policy* 32: 630-634.

Greboval, D. E. 2002. *Report and documentation of the international workshop on factors contributing to unsustainability and overexploitation in fisheries*, p. 12pp. Rep. 672, Rome, Italy.

---. 2004. *International workshop on the implementation of international fisheries instruments and factors of unsustainability and overexploitation in fisheries.* Rep. 700, Rome, Italy, 72pp.

Guidetti, P., and J. Claudet. 2010. Comanagement practices enhance fisheries in marine protected areas. *Conservation Biology* 24: 312-318.

Halpern, B. S., S. D. Gaines, and R. R. Warner. 2004. Confounding effects of the export of production and the displacement of fishing effort from marine reserves. *Ecological Applications* 14: 1248-1256.

Halpern, B. S., K. A. Selkoe, F. Micheli, and C. V. Kappel. 2007. Evaluating and ranking the vulnerability of global marine ecosystems to anthropogenic threats. *Conservation Biology* 21: 1301-1315.

Harmelin-Vivien, M. and others 2008. Gradients of abundance and biomass across reserve boundaries in six Mediterranean marine protected areas: evidence of fish spillover? *Biological Conservation* 141: 1829-1839.

Hay, M. E., and P. R. Taylor. 1985. Competition between herbivorous fishes and urchins on Caribbean reefs. *Oecologia* 65: 591-598.

Hilborn, R. 2007a. Managing fisheries is managing people: what has been learned? *Fish and Fisheries* 8: 285-296.

---. 2007b. Moving to sustainability by learning from successful fisheries. *Ambio* 36: 296-303.

Hind, E. J., M. C. Hiponia, and T. S. Gray. 2010. From community-based to centralised national management-A wrong turning for the governance of the marine protected area in Apo Island, Philippines? *Marine Policy* 34: 54-62.

ICRAN. 2002. People and reefs: a partnership for prosperity. *International Coral Reef Action Network.*

ICRI. 2002. Deciding on priorities to protect and manage the world's coral reefs. *International Coral Reef Initiative.*

Jennings, S., S. S. Marshall, and N. V. C. Polunin. 1996. Seychelles' marine protected areas: comparative structure and status of reef fish communities. *Biological Conservation* 75: 201-209.

Johannes, R. E. 2002. The renaissance of community-based marine resource management in Oceania. *Annual Review of Ecology and Systematics* 33: 317-340.

Johannes, R. E., and M. Riepen. 1995. *Environmental, Economic, and Social Implications of the Live Reef Fish Trade in Asia and the Western Pacific.*

Jones, R. J., and A. L. Steven. 1997. Effects of cyanide on corals in relation to cyanide fishing on reefs. *Marine and Freshwater Research* 48: 517-522.

Kelleher, G., C. Bleakley, and S. Wells. 1995. *A global representative system of marine protected areas.* The World Bank.

Kelleher, G., and C. Recchia. 1998. Lessons from marine protected areas around the world. *Parks* 8: 1-4.

King, M., and U. Faasili. 1999. Community-based management of subsistence fisheries in Samoa. *Fisheries Management and Ecology* 6: 133-144.

Kritzer, J. P. 2004. Effects of noncompliance on the success of alternative designs of marine protected-area networks for conservation and fisheries management. *Conservation Biology* 17: 630-632.

Larkham, A. W. D. 1983. The primary productivity of plant communities, p. 221-230. *In* D. J. Barnes [ed.], *Perspectives on coral reefs. AIMS.*

Lewis, J. B. 1977. Processes of organic production on coral reefs. *Biological Review* 52: 305-347.

Liu, P. L. F. and others 2005. Observations by the international tsunami survey team in Sri Lanka. *Science* 308: 1595.

Lucas, J. S. 1994. The biology, exploitation, and mariculture of giant clams (Tridacnidae). *Reviews in Fisheries Science* 2: 181-224.

Mascia, M. B. 2003. The human dimension of coral reef marine protected areas: Recent social science research and its policy implications. *Conservation Biology* 17: 630-632.

Matthew, S. 2001. Small-scale fisheries perspectives on an ecosystem-based approach to fisheries management, p. 18. *Responsible fisheries in the marine ecosystem,* FAO.

McClanahan, T. R. 1995. A coral reefs ecosystem-fisheries model: impacts of fishing intensity and catch selection on reef structure and processes. *Ecological Modelling* 80: 1-19.

---. 1999. Is there a future for coral reef parks in poor tropical countries? *Coral Reefs* 18: 321-325.

McClanahan, T. R., V. Hendricks, M. J. Rodriques, and N. V. C. Polunin. 1999a. Varying responses of herbivorous and invertebrate-feeding fishes to macroalgal reduction on a coral reef. *Coral Reefs* 18: 195-203.

McClanahan, T. R., C. C. Hicks, and E. S. Darling. 2008. Malthusian overfishing and efforts to overcome it on Kenyan coral reefs. *Ecological Applications* 18: 1516-1529.

McClanahan, T. R., and S. C. Mangi. 2004. Gear-based management of a tropical artisanal fishery based on species selectivity and capture size. *Fisheries Management and Ecology* 11: 52-60.

McClanahan, T. R., N. A. Muthiga, A. T. Kamukuru, H. Machano, and R. W. Kiambo. 1999b. The effects of marine parks and fishing on coral reefs of Northern Tanzania. *Biological Conservation* 89: 161-182.

McClanahan, T. R., and D. Obura. 1995. Status of Kenyan coral reefs. *Coastal Management* 23: 57-76.

---. 1997. Sedimentation effects on shallow coral communities in Kenya. *Journal of Experimental Marine Biology and Ecology* 209: 103-122.

McClanahan, T. R., N. Polunin, and T. Done. 2002. Ecological states and the resilience of coral reefs. *Conservation Biology* 6: 18.

McClanahan, T. R., and S. H. Shafir. 1990. Causes and consequences of sea urchin abundance and diversity in Kenyan coral reef lagoons. *Oecologia* 83: 362-370.

McCook, L., J. Jompa, and G. Diaz-Pulido. 2001. Competition between corals and algae on coral reefs: a review of evidence and mechanisms. *Coral Reefs* 19: 400-417.

McManus, J. W. 1997. Tropical marine fisheries and the future of coral reefs: a brief review with emphasis on Southeast Asia. *Coral Reefs* 16: S121-S127.

Mora, C. and others 2006. Coral reefs and the global network of marine protected areas. *Science* 312: 1750-1751.

Mous, P. J., L. Pet-Soede, M. Erdmann, H. S. J. Cesar, Y. Sadovy, and J. S. Pet. 2000. Cyanide Fishing on Indonesian Coral Reefs for the Live Food Fish Market – What is the Problem? *SPC Live Reef Fish Information Bulletin* 7: 20-27.

Mumby, P. J. 2006. Connectivity of reef fish between mangroves and coral reefs: Algorithms for the design of marine reserves at seascape scales. *Biological Conservation* 128: 215-222.

Mumby, P. J., and A. R. Harborne. 2010. Marine reserves enhance the recovery of corals on Caribbean reefs. *PLoS ONE* 5: 1-7.

Mumby, P. J., A. Hastings, and H. J. Edwards. 2007. Thresholds and the resilience of Caribbean coral reefs. *Nature* 450: 98-101.

Munro, J. L. 1989. Fisheries for giant clams (Tridacnidae: Bivalvia) and prospects for stock enhancement, p. 558. In J. F. Caddy [ed.], *Marine invertebrate fisheries: their assessment and management*. John Wiley & Sons.

Muthinga, N. A., S. Riedmiller, R. Van Der Elst, J. Mann-Lang, C. Horrill, and T. R. McClanahan. 2000. Management status and case studies. *In* T. R. McClanahan, C. S. Sheppard and D. O. Obura [eds.], *Coral reefs of the Indian Ocean: their ecology and conservation.* Oxford University Press.

Nagelkerken, I. and others 2002. How important are mangroves and seagrass beds for coral-reef fish? The nursery hypothesis tested on an island scale. *Marine Ecology-Progress Series* 244: 299-305.

Nagelkerken, I., G. Van Der Velde, M. W. Gorissen, G. J. Meijer, T. Van't Hof, and C. Den Hartog. 2000. Importance of mangroves, seagrass beds and the shallow coral reef as a nursery for important coral reef fishes, using a visual census technique. *Estuarine Coastal and Shelf Science* 51: 31-44.

Nasuchon, N., and A. Charles. 2010. Community involvement in fisheries management: Experiences in the Gulf of Thailand countries. *Marine Policy* 34: 163-169.

Okello, M. M., and J. W. Kiringe. 2004. Threats to biodiversity and their implications in protected and adjacent dispersal areas of Kenya. *Journal for Sustainable Tourism* 12: 55-69.

Palumbi, S. R. 2004. Marine reserves and ocean neighborhoods: the spatial scale of marine populations and their management. *Annual Review of Environment and Resources* 29: 31-68.

Patterson, T., T. Gulden, K. Cousins, and E. Kraev. 2004. Integrating environmental, social and economic systems: a dynamic model of tourism in Dominica. *Ecological Modelling* 175: 121-136.

Pauly, D. 1988. Some definitions of overfishing relevant to coastal zone management in Southeast Asia. *Tropical Coastal Area Management* 3: 14-15.

---. 1990. On Malthusian overfishing. *Naga* 13: 3-4.

---. 1994. From Growth to Malthusian Overfishing: Stages of Fisheries Resource Misuse. *Traditional Marine Resource Management Knowledge Bulletin* 3: 7-14.

Pauly, D. and others 2002. Towards sustainability in world fisheries. Nature 418: 689-695.

Pauly, D., G. Silvestre, and I. R. Smith. 1989. On Development, Fisheries and Dynamite: A Brief Review of Tropical Fisheries Management. *Natural Resource Modelling* 3: 307-329.

Pendleton, L. H. 1995. Valuing coral reef protection. *Ocean & Coastal Management* 26: 119-131.

Pet-Soede, C., H. S. J. Cesar, and J. S. Pet. 1999. An economic analysis of blast fishing on Indonesian coral reefs. *Environmental Conservation* 26: 83-93.

Pet-Soede, L., and M. Erdmann. 2003. *Rapid Ecological Assessment of the Wakatobi National Park*. Marine Program, WWF Indonesia.

Pilcher, N. J., and S. G. Oakley. 1997. Unsustainable fishing practices: crisis in coral reef ecosystems in south east Asia., p. 77-87. *Oceanology International* 97 Pacific Rim, 12-14 May 1997.

Pollnac, R. B., B. R. Crawford, and M. L. G. Gorospe. 2001a. Discovering factors that influence the success of community-based marine protected areas in the Visayas, Philippines. *Ocean and Coastal Management* 44: 683-710.

Pollnac, R. B., and J. Poggie. 2006. Job satisfaction in the fishery in two southeast Alaskan towns. *Human Organization* 65: 332-342.

Pollnac, R. B., R. S. Pomeroy, and I. H. T. Harkes. 2001b. Fishery policy and job satisfaction in three poverty traps. *Journal of Econometrics* 77: 5-37.

Pomeroy, R., and F. Berkes. 1997. Two to tango: the role of government in fisheries co-management. *Marine Policy* 21: 465-480.

Pomeroy, R. S. 1995. Community-based and co-management institutions for sustainable coastal fisheries management in Southeast Asia. *Ocean and Coastal Management* 27: 143-162.

Pomeroy, R. S., B. M. Katon, and I. Harkes. 2001. Conditions affecting the success of fisheries co-management: lessons from Asia. *Marine Policy* 25: 197-208.

Pretty, J., and D. Smith. 2004. Social capital in biodiversity conservation and management. *Conservation Biology* 18: 631-638.

Randall, A. 1991. The value of biodiversity. Ambio 20: 64-67.

Resosudarmo, B. P. 2005. *The politics and economics of Indonesia's natural resources*. RFF Press.

Richardson, E. A., M. J. Kaiser, G. Edwards-Jones, and H. P. Possingham. 2006. Sensitivity of marine-reserve design to the spatial resolution of socioeconomic data. *Conservation Biology* 20: 1191-1202.

Robbins, P. 2000. The rotten institution: corruption in natural resource management. *Political Geography* 19: 423-443.

Roberts, C. M. 1997. Ecological Advice for the Global Fisheries Crisis. *Trends in Ecology and Evolution* 12: 35-38.

Roberts, C. M., J. A. Bohnsack, F. Gell, J. P. Hawkins, and R. Goodridge. 2001. Effects of marine reserves on adjacent fisheries. *Science* 294: 1920-1923.

Roberts, C. M., J. P. Hawkins, and F. R. Gell. 2005. The role of marine reserves in achieving sustainable fisheries. *Philosophical Transcriptions of the Royal Society B* 360.

Roberts, C. M., and N. V. C. Polunin. 1993. Marine Reserves - Simple Solutions to Managing Complex Fisheries. *Ambio* 22: 363-368.

Rudd, M. A., M. H. Tupper, H. Folmer, and G. C. Van Kooten. 2003. Policy analysis for tropical marine reserves: challenges and directions. *Fish and Fisheries* 4: 65-85.

Ruitenbeek, J., and C. Cartier. 1999. *Marine system valuation: an application to coral reef systems in the developing tropics,* p. 149. *In* R. Huber [ed.], Issues in applied coral reef biodiversity valuation: results for Montego Bay, Jamaica. World Bank.

Ruitenbeek, J., R. M. Huber, and K. Gustavson. 2000. *Integrated coastal zone management of coral reefs: decision support modelling*, Volume 1. World Bank.

Sale, P. F. and others 2005. Critical science gaps impede use of no-take fishery reserves. *Trends in Ecology and Evolution* 20: 74-80.

Salm, R. V., and M. Ngoile. 1998. Marine protected areas: changes and challenges. *In* R. V. Salm and Y. Tessema [eds.], Partnership for conservation: report of the regional workshop on marine protected areas, tourism and communities, Diani Beach, Kenya, May 11-13 1998. IUCN Eastern Africa Regional Office.

Schiel, D. R., and W. A. Nelson. 2004. The harvesting of macroalgae in New Zealand. *Hydrobiologia* 204-205: 25-33.

Sievanen, L., B. Crawford, R. Pollnac, and C. Lowe. 2005. Weeding through the assumptions of livelihood approaches in ICM: seaweed farming in the Philippines and Indonesia. *Ocean & Coastal Management* 48: 297-313.

Smith, D. J., S. E. Pilgrim, and L. C. Cullen. 2007. Coral Reefs and People. In J. Pretty et al. [eds.], *Sage Handbook on Environment and Society*. Sage Publications.

Spalding, M. D., C. Ravilious, and E. P. Green. 2001. *World atlas of coral reefs*. UNEP World Conservation Monitoring Centre.

Spurgeon, J. P. G. 1992. The economic valuation of coral reefs. *Marine Pollution Bulletin* 24: 529-536.

Stefansson, G., and A. A. Rosenberg. 2005. Combining control measures for managing fisheries under uncertainty: quotas, effort limitation and protected areas. *Philosophical Transcriptions of the Royal Society B* 360: 133-146.

Suvapepun, S. 1994. *Environmental impacts of mariculture*. Chulalongkorn University.

Tamayo, G., W. F. Nader, and A. Sittenfield. 1997. *Biodiversity for bioindustries*. In B. V. Ford-Lloyd, H. J. Newbury and J. A. Callow [eds.], *Biotechnology and plant genetic resources: conservation and use*. CAB International.

Tissot, B. N., W. J. Walsh, and M. A. Hixon. 2009. Hawaiian Islands Marine Ecosystem Case Study: Ecosystem- and Community-Based Management in Hawaii. *Coastal Management* 37: 255-273.

Unsworth, R. K. F., and L. C. Cullen. 2010. Recognising the necessity for Indo-Pacific seagrass conservation. *Conservation Letters* 3: 63-73.

Unsworth, R. K. F., P. S. De Leon, S. L. Garrard, J. Jompa, D. J. Smith, and J. J. Bell. 2008. High connectivity of Indo-Pacific seagrass fish assemblages with mangrove and coral reef habitats. *Marine Ecology-Progress Series* 353: 213-224.

Unsworth, R. K. F., J. D. Taylor, A. Powell, J. J. Bell, and D. J. Smith. 2007a. The contribution of scarid herbivory to seagrass ecosystem dynamics in the Indo-Pacific. *Estuarine Coastal and Shelf Science* 74: 53-62.

Unsworth, R. K. F., E. Wylie, D. J. Smith, and J. J. Bell. 2007b. Diel trophic structuring of seagrass bed fish assemblages in the Wakatobi Marine National Park, Indonesia. *Estuarine Coastal and Shelf Science* 72: 81-88.

Waycott, M. and others 2009. Accelerating loss of seagrasses across the globe threatens coastal ecosystems. *Proceedings of the National Academy of Sciences* 106: 12377-12381.

Weis, J. S., and P. Weis. 2005. Use of Intertidal Mangrove and Sea Wall Habitats by Coral Reef Fishes in the Wakatobi National Park, Indonesia. *Bulletin of Zoology* 53: 119-124.

White, A. T., V. Barker, and G. Tantrigama. 1997. Using integrated coastal management and economics to conserve tourism resources in Sri Lanka. *Ambio* 26: 335-344.

White, A. T., A. Salamanca, and C. A. Courtney. 2002. Experience with marine protected area planning and management in the Philippines. *Coastal Management* 30: 1-26.

Whittingham, E., J. Campbell, and P. Townsley. 2003. *Poverty and reefs: a global overview*. Intergovernmental Oceanographic Commission of UNESCO.

Wilen, J. E., M. D. Smith, D. Lockwood, and L. W. Botsford. 2002. Avoiding surprises: incorporating fishermen behavior into management models. *Bulletin of Marine Science* 70: 553-575.

Wilkinson, C. R. 2008. Status of coral reefs of the world: 2008, p. 296pp. *Global Coral Reef Monitoring Network and Reef and Rainforest Research Center*.

Wright, A. 1985. *Marine resource use in Papua New Guinea: can traditional concepts and contemporary development be integrated? In* K. Ruddle and R. Johannes [eds.], The traditional knowledge and management of coastal systems in Asia and the Pacific. UNESCO.

In: Fishery Management
Editor: Janice S. Intilli

ISBN: 978-1-61209-682-7

Chapter 2

REGULATORY TOOLS AND FISHERIES MANAGEMENT: AN EVALUATION BASED ON INLAND CAPTURE FISHERIES OF BANGLADESH

Abdullah-Al Mamun*
Natural Resources Institute, University of Manitoba,
303-70 Dysart Road, Winnipeg, MN R3T 2N2, Canada

ABSTRACT

This paper focuses on the regulatory tools and related mesaures (e.g., acts, rules, policies and plans) of Bangladesh that define the management and conservation of inland capture fisheries in the rivers, floodplains and *beel lakes* of Bangladesh. The goal of this paper is to help policy-makers and scientific communities to rethink conservation policies and help them realize that in implementing and developing policy local level participation in decision-making and their livelihoods issues are vital. This study recognizes that the natural resource policy tools of Bangladesh are very broad and have long history of managing natural resources but with poor enforcement. Fisheries management also lacks participatory dialogue during the policy development process. The policy tools, therefore, are not very effective and so lead to poor compliance and impede the sustainable development of inland fisheries. Given the apparent inadequacy of the existing management and regulatory tools, this paper argues for multi-stakeholder decision making involving fishers and other local stakeholders at the policy formation stage which can lead to better management of inland capture fisheries.

Keywords: Inland fisheries, Regulatory tools, Habitat degradation, Participatory decision-making

* Department of Geography and Environmental Studies, Wilfred Laurier University, 75 University Avenue West, Waterloo, ON N2L 3C5, Canada. e-mail: mamu6650@wlu.ca

1. Introduction

Everyday in Bangladesh and around the world the lives of millions of people involve the day-to-day use of natural resources such as fisheries. Maintaining a balance between healthy resource-use systems and resource stocks is vital for maintaining social, ecological and economic functions. This is often the major focus for management decisions seeking sustainable use of common pool resources. Achieving such a goal, however, is a complex process and not easy especially for developing countries that struggle with limited financial resources and low institutional capacity. These obstacles have perhaps created awareness for developing more innovative technical, social and political institutions such as co-management and integrated management to ensure livelihoods security through saving depleted resources. But in reality many resource users never have the effective opportunity to contribute to the development of policies and institutions that govern the use of the natural resources upon which their livelihoods depend (IIED 2010; Unhoff 1992). The policy gap between the powerful (resource managers) and the marginalized (resource users) often pertains to a lack of opportunities for participatory dialogue and identifying areas of discussions and related processes for joint decision making. Even when such opportunities or channels exist, communication may fail due to fundamental differences in understanding, perceptions, and expressions among stakeholders and the different power relationships between user groups and policy-makers. Given this understanding, this paper describes fisheries management tools (acts, ordinances and rules etc.), the the roles of actors involved in making such rules and identifies the effectiveness of past and present policy tools. Secondly, this paper looks at how perceptions between fishers and policy-makers vary on some important fisheries management decisions (e.g., establishing connectivity for a degraded fisheries and habitat enhancement) and enforcing regulations such as Fish Acts. Finally, this paper introduces participatory governance paradigms such as co-management as options for creating dialogues between stakeholders for developing a more acceptable and effective policy formulation process. This approach may be helpful because both fishers' opinions as well as scientific understanding of managers can be incorporated into the policy process. If the knowledge of resources users is effectively incorporated into policy processes it can lead to increased local legitimacy and thus better rule compliance.

This chapter focuses on the inland open water fisheries of Bangladesh that are used extensively by both full-time and subsistence fishers and are often considered as a degraded resource. Like two other fisheries sectors (inland aquaculture and marine fisheries), inland capture fisheries contribute greatly to the incomes of a large number of people. In Bangladesh, they contribute to the incomes of more than 22 million rural people (FAO 2010). Inland fisheries is also a inexpensive source for protein, micro-nutrients, fatty acids and minerals for the rural poor (Roos and Thilsted 2003). However, the inland aquatic ecosystem has become highly altered due to human uses such as overfishing, cross-dams and the agricultural uses of fishing areas. Globally, open water inland fisheries are continually degraded but yet still make vitally important contributions to many communities in Asia and Africa (FAO 2007). The assessment and evaluation made in this chapter related to fisheries policy and management processes and the recommendations made here may also apply to other regions of the world as well given the worldwide degrading trends of inland fisheries.

1.1. Inland Fisheries: Biophysical and Socio-Economic Features

Bangladesh is formed from deltas and has number small (Ismoty, Gorai, Kobatak, Korotoa, and Boral etc.) and large river systems (e.g., *Padma, Meghna and Jamuna*). Seasonal flood waters from rainfall orginiate in upper reaches of these river systems and subsequently inundate an average 3 million ha of the country each year from June to October (Thompson 2006). The floodplain includes natural lakes called either *beels* and *haors* depending on location, size and uniqueness to fish production. Open water fisheries contain high species and size diversity (Mamun 2006) which differs from aquaculture ponds (e.g., monoculture of carps). These diversified open water habitats provide important opportunities for fishing with minimum or no investments except primarily labor and traditional gears and traps. As such open water fisheries have been very important in supporting the subsistence economy in rural Bangladesh by providing readily available fish to sell or to daily consumption. The inland open waters are inhabited by 260 species of fish and 25 species of shrimp (Mamun 2006). In the 1960's, the inland open water fishery resources contribute approximately 90 per cent of the country's fish production (FAO 2010). However, it is generally agreed that there has been considerable reduction of the total fish catch from open water fisheries in inland waters (See Thomson 2006)). Man-made causes, such as overfishing and lack of better management measures and inadequate policies to address production in the inland open waters, are blamed for this declining trends.

Inland water systems also include dugout ponds for carp cultivation and other fishes (mostly catfishes, *Pangasius hypophthulmus)*. The growth of fisheries production from aquaculture and the marine sector has greatly increased in recent decades with the production from ponds almost exceeding the total production from the open water capture fisheries (see Table 1). Pond aquaculture is facilitated by easy and available technology (Hussain 2010). In 1984-85, the total catch from inland open waters was 194,130 mt which increased to 956,686 mt in 2005-06. However, this increased catch is from culture-based fisheries in *baors* (oxbow lakes) and some *beels* stocked with carp fingerings (Valbo-Jørgensen and Thompson 2007). Over recent decades coastal areas of Southern Bangladesh have experienced increased level of shrimp farm expansion. Coastal shrimp culture has also helped in the overall increase in fish production from aquaculture but at the cost of environmental degradation such as loss of coastal mangrove forests (SEHD 2003)].

Table 1. Area wise production of fishes from different water bodies of Bangladesh in 2005- 2006 (Source FAO 2007).

Water body category	Area	Total catch (Ton)	Catch/Area (kg/ha)
Rivers and estuarine	1 031 563	137 859	134
Sunbdarban	*	16 423	-
Beels/haors	114 161	74 365	669
Kaptai lake	68 800	7 548	110
Floodplains	2 832 792	718 491	254
Total Inland open waters	4 047 316 (49%)	956686(41%)	669
Inland Aquaculture (Ponds, ditches, shrimp farms etc.)	528 390 (6%))	892049 (38%)	1688
Marine (Artisanal and off-shore fisheries)	**7063069 (45%)	472688.2 (20%)	12
Total production		2321423.2(100%)	-

(* Catch from Sundarban is included in rivers and estuaries; ** Considers only 50 m depth area used as fishing zone)

Intensive aquaculture production in Bangladesh is partly driven my market economy and benefits people who own the necessary land for excavation of ponds and resources to operate such as rearing costs. Small fish, on the other hand, come mostly from open water capture fisheries. These small fish are caught from the floodplains by poor people and are neglected in official statistics and policies (Thomson 2006). Small fishes, however, are the accessible and preferred food of poor people and are good sources of micro-nutrients (Roos and Thilsted 2004). In recent decades managing these small fishes in open waters that benefit local livelihoods has been an important policy decision as these fish are declining from inland systems.

There is no reliable statistics related to the total number of fishers engaged in fisheries. However, a conservative estimate says that fisheries secotor provide employment for 770 000 fishers through access to the water bodies (FAO 2007). Over 167 000 fishers and support staff are employed in the marine sector. In addition, about 185 000 people are engaged in part-time shrimp fry collection activities at the coast of southern Bangladesh. Most of the rural population fish professionally, seasonally or for subsistence. Small to medium size fishers provide the majority of the national open water fish catch. In addition to the harvesters, a further two million people are involved in activities related to the fisheries sector such as fish trading, drying and transportation. Although fish catch comes from marine sources too, the level of human interactions with inland waters are more intense with respect to human use of inland waters (e.g., agricultures, roads etc.) that have caused various damaged to fisheries and for this paper considers inland fisheries.

2. Research Methods and Data Collection

This research contains both primary and secondary data from a project (2006-2007) on fisher ecological knowledge on floodplain systems of Bangladesh (Mamun 2007) as well as a subsequent desktop search inventing fisheries regulatory tools of Bangladesh. Primary data set (e.g. fisher opinion to fisheries management and response of fishers to management policy) was collected from a *beel* system (a natural lake) composed of three water bodies; *Chakidam, Kuradanga* and *Solmara* (see figure 1). The study is typical of degraded inland fisheries and has several problems that are very common to inland fisheries of Bangladesh. For example, migratory routes for fishes were found silted; many parts of the beel are altered for other uses (e.g. agricultures), and traditional fishers are very dependent of on wild fish and aquatic products such a fodder for cattle feeding from the beels. Importantly, government conservation and management policies have had little positive impact on the fisheries of the study area. There is a considerable reduction in natural fish production as compared to earlier catch volume as perceived by fisheries.

Data collection on the impact of regulatory tools on fisheries management and livelihoods of fishers and local community included two focus groups discussion (5-10 participants), as well as personal interviews to obtain insight on the impacts of policy tools (such as Fish Acts) on conservation and the overall directions of fisheries management.Total number of interviewees was 50 which includes community fishers (N=25), community leaders (N=2), farmers (N=10), fish traders (N=5), fisheries managers (N=4), researchers (N=2) and NGO officials (N=2),. In addition, secondary data was also used to review past policies and regulatory elements applicable to inland fisheries and water management.

Academic and grey literature from recent co-management and integrated water management sources such as Department of Fisheries (DoF), Bangladesh, Bangladesh Water Development Board and WorldFish Center, Dhaka Office are used as secondary sources of data.

Many of the conclusions, with respect to directing a new and refined policy for guiding future fisheries management are products of several personal meetings and interviews with fishers, community leaders and fisheries managers.

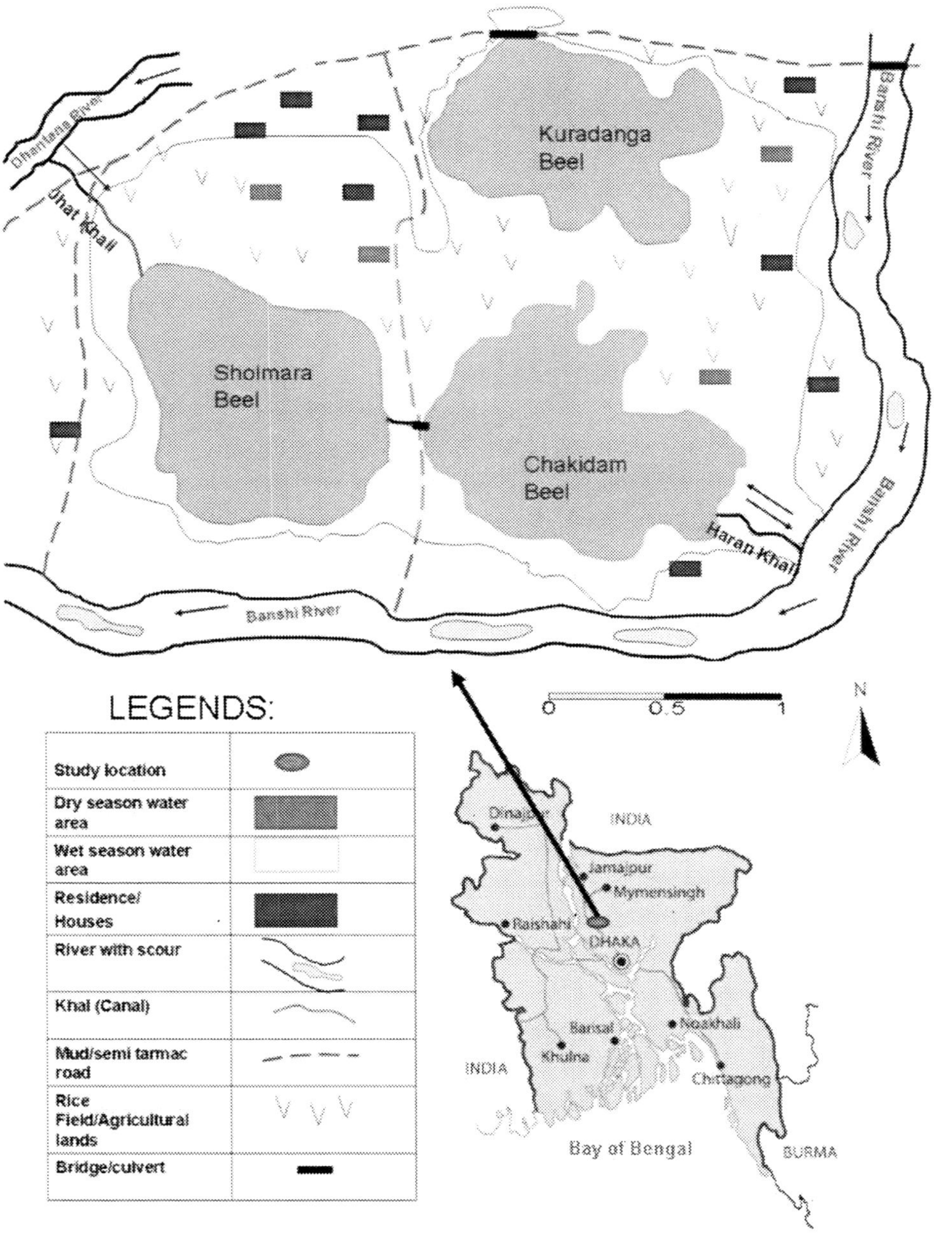

Figure 1. Sketch map of the study area

3. RESULTS AND DISCUSSIONS

3.1. Overview of the Legal and Policy Frameworks and Accesses

The previous subsections establish fisheries as an important sector in Bangladesh for free protein, easy incomes for rural people and contributions to national economy. As such ensuring sustainable management of this sector is particularly important in the light of current degraded state of inland fisheries (SEHD 2003). The management of fisheries in Bangladesh is guided by existing fisheries acts, rules and polices that defines the access to fisheries, rights and related conservation measures. As a result, the existing access rights, legal and policy tools and their enforcement mechanisms are important to consider with respect to understanding options for management tools. Given the nature of property rights and harvesting processes, wild fisheries in Bangladesh can be considered as common property especially in the wet season (July-September). During the wet season very limited restrictions are applied with respect to access to any type of water bodies including waters on private lands. During the dry season (January-May), however, access is partially restricted in many fishing grounds, especially where water bodies are leased out to individuals or where fishing is confined to private lands such as the *beels* in this study (see Figure1).

The study area consists of 95% private lands but it was found that fishing is not restricted on such lands during wet season. Informal institutional arrangements between land owners and fishers prevail that allows everybody to fish. For example, sometimes fishers happily give a small portion of their catch to nearby landowners. Such informal management institution prevails in floodplain fisheries all over the country and has continued historically and has probably served to minimize conflicts between fishers and land owners.

In parallel with traditional informal management of inland fisheries, the government of Bangladesh has always relied on regulatory instruments. A list of regulatory instruments applicable to fisheries is given in Table 2. There are some floodplain and closed areas where the government stocks carp fingerlings to supplement natural breeding of fish. For both inland capture fisheries and culture-based capture fisheries fisheries in open waters, the role of policy tools are equally important as they provide important management directions to fishing access rights, fishing seasons, harvestable sizes and gears to be used.

The Tank Improvement Act 1939 could be cosideded as the fisrt regulatory instruments in inland waters that describes the water use rights by private individuals on the basis of certain amount of tax paid to the government. However, this Act does not say much about management of fisheries but rather talks about fish harvesting rights.

The second key instrument is The Protection and Conservation of Fish Act 1950. For many reasons, this Act is very powerful as it includes the major issue of fisheries degradation such as catching under-sized fishe and the prohibition of those fixed gears which obstruct fish migration routes. Fish Act 1950 outlines how to conserve fish habitats against activities such as flood control, drainage and irrigation (FCD/I), agricultural, industries, roads and urban development projects (Section 6.1 of the policy). It also states in the act that the breeding grounds of fish should be conserved (Section 6.5 of the policy) but does not say much about the exact strategy to enforce the Act and the implementation process.

This Act is powerful in the sense that it still is considered as major regulatory measure for inland open water fisheries and many of the clauses from it have transferred to other successive acts and policies such as Fish Policy 1998. Based on Fish Act 1950 A surveillance to monitor compliance to closed season (May-August) is carried out in floodplains and rivers by teams operating under district level fisheries officers and assigned magistrates. However, it was found that 90% of the fishers and local stakeholders interviewed in this study are not very aware of the Fish Acts and their implementation. They also did not confirm any visits in the past by fisheries departement to implement Fish Act in their fisheries area.

In addition, DoF mostly can not exercise their legal authority to enforce the Fish Act 1950 on their own. For example, confiscating gear for violation would have to be in the presence of judiciary personnel such as sub-district level magistrates. However, seeking such helps from district level is a complex and time consuming process. Therefore, the enforcement process often leads to inefficient measures during the closed season (May-August). This is the peak breeding season of carps and some other commercial fish such as catfishes (*Mystus spp.*)

Other limitations of this Fish Act is its major focus on the management of a few local major carps (rui *Labeo rohita,* catla *Catla catla* and mrigal *Cirrhinus cirrhosus*), river shad (hilsha *Tenualosa ilisha*) fisheries and several other catfishes (e.g. ayre *Sperata aor*) and snakeheads (e.g. Shol/Taki *Channa* sp.) of commercial importance. Interesting to note that currently carps do not contribute much to the open water catch (less then 5% only) (FAP-17 1994). A recent analysis of catch trends from WorldFish records including two river sections and one semi-closed water body and one closed water body also indicates the same trends (see Table 3). It was found that overall carp catches from all five types of water bodies ranges from 1-5% and the rest is made up of other small fishes. It is a major limitation that this Act and most of the management interventions are directed towards conserving carp fisheries and neglects principles and guidelines for conserving indigenous species found in inland waters (Thompson et al. 2000). The management requirements for small species, however, would be certainly different from the Indian major carps as many indigenous species are non-migratory and opportunistic breeders. They are capable of multiple spawnings and in a diversity of habitats (FAP-17 1994).

Table 3. Percentage of carp catches (by weight) from 5 water bodies (1997-2002).

Name and type of water body	1997	1998	1999	2000	2001	2002
Dikshi Beel (closed)	0.8	9.3	6.9	3.9	1.1	2.9
Asur beel (semi-closed)	0.5	0	3.4	0	0	0.3
kali River (flowing river)	1.3	1.5	0.8	1.5	0.9	0.7
Titas River(flowing river)	11.5	11.5	6	6.2	2.5	9.9
Overall carps (Rui, Katol, Mrigal etc.)	4	5.5	4	2.9	1.1	3.4

(Source: WFC 2006)

Conserving these fisheries would require establishing dry season refuges. Prohibition dry pumping of small ponds or ditches or other deeper areas of water bodies would also be important. At the present, Fish Act 1950 is only enforced during the wet season (July-September) ensuring breeding of fish but have no provisions for dry season conservation. Also a new rule may be required to discourage draining fisheries areas for dry season irrigation. It is evident that the Act of 1950 and its subsequent version the Protection and Conservation of Fish (Amendment) Act 1995 hardly meets the present day needs of fisheries management (Ali and Thompson 2006; Hughes et al. 1994; Ali 1997; Minkin and Boyce 1994). As a result the declining trends in fish catch and increased poverty of fishers have put contemporary management in question (Minkin and Boyce 1994).

Table 2. Summary of existing acts ordinances and regulations fisheries in Bangladesh.

Periods	Name of policy and management systems	General management functions/objectives	Observations on achievements and outcomes/comments
1. 1. Pre-colonial (before 1757)	1.1 No report of management measures but informal institutions	Exploit measure to harvest locally managed though community norms and principles.	Presumably less fishers and no scares resource
2. Colonial and after and pre-independence (1757-1971)	2.1 Feudal lords (*zamindars*) appointed by colonial rulers to govern lands	The collection of tax in exchange for user rights by feudal lords on behalf of the colonial ruler.	Limited public access and for commercial uses.
	2.2 The Tank Improvement Act, 1939 State Land Acquisition and Tenancy Act, 1950	Lords assigned for authorization to fishing and use Tanks. End of the Feudal lords system. Decision related to collecting land rent paid off by the tenants to government.	No provision management of capture fisheries. Ministry of Land (MOL) obtained property rights for public water bodies. Greater impacts on ceasing communal rights on fishing.
	2.3 East Bengal protection and conservation of Fish Act 1950	Prohibition of undersized nets and sanctions and regulation of destructive use of chemicals for fishing and pesticides in fisheries area	A powerful act with respect to its coverage. Weak implementation.
3. Post-independent (1971 till date)	3.1 The protection and conservation of fish (amendment) ordinance, 1982	Contains several amendments to promote the conservation measures laid out at Fish Act 1950.	Fuller implementation is yet to realize.
	3.2 The marine Fishers ordinance, 1983	Directed to marine and costal area to set license management.	Declaration fish reserve/banning on fishing areas for *hilsha*.
	3.3. Environment Policy, 1992	Focuses ecological restoration environmental balance.	Weak enforcement similar fate as Fish Act 1950.
	3.4 The New Fisheries Management Policy (NFMP) 1986	Progressive towards re-establishing rights of genuine fishers.	Limited impacts for influential lease-holders.
	3.5 Water Resource Planning Act 1992	Balanced use of water resources	Progressive for inclusion of lower level of admin.
	3.6 Bangladesh Environment Conservation Act, 1995	Very detail policy given environment is a broader issue. Avoidance of environmental damage.	Weak enforcement as no sanctioning under this act effective.
	3.7 The National Fish Policy 1998	Organize genuine fishers Gradual abolition of leasing with gear-wise licensing. Biological management and socio- economic improvement covers all types of fisheries.	Very details and ambitious in the context of complex management problems and fear of poor implementation.

The third set of instruments in Bangladesh that affect fisheries management is the State Acquisition and Land Tenancy Act, 1950 and a subsequent policy that triggered the process of leasing aquatic areas to fisheries cooperatives called the New Fisheries Management Policy (NFMP). The Land Act serves to lease out water bodies to private operators as all open and some closed waters that are not private properties are public lands. Under the Land Tenancy Act any tenable fisheries areas are called *Jalmohals (*water states) and are leases yearly by the government to highest bidders. The leasing of *Jalmohals* is an older practice and started after independence in 1971 under pressure to develop and use resources to meet management costs of the government. At present there are about 14,000 government designated *Jalmohal* areas in Bangladesh contributing 0.7% of GDP (Thompson 2006). This approach bought the market into open water fisheries' system as water bodies are privatized though open auctions/bids. This policy brought incomes for the government and saw some success in preventing illegal land grabbing and the conversion of many wetland areas into agricultural lands. However, this approach has achieved limited success in preventing habitat damage and reducing excessive fishing pressure, major concerns during its operational stage (1971-1986). Secondly, under *Jalmohal* policy no conservation rules are applied to restricting illegal fishing even during fish breeding seasons as the area is private. As well, The *Jalmahal* policy does not really support traditional informal institutional arrangements (community-managed fishing) which effectively abolished by the government under the *Jalmohal* system. It is argued by Toufique (2006) that the *Jalmohal* management system favored mainly outsiders as they could bid on high lease prices and thus gain authority over access to fishing. On the other side, the genuine fishers who are mostly poor could not participate in the bidding process due to their financial constraints. In some cases, fishers are involved in *Jalmohal* fisheries, but only as day laborers. As a result, there are no benefit sharing mechanisms between leaseholders and the wider fishing communities. Given these negative outcomes the *Jalmohal* system was suspended in 1995 especially in many rivers.

The *Jalmahal* policy, however, was reorganized later by introducing a new policy called the New Fisheries Management Policy (NFMP) 1986. It was perceived that The NFMP would be a way to prevent the exploitation of genuine fishermen by leaseholders and to re-establish communal fisheries rights. This would enhance the ability of permanent fishers (mostly from a cast of hindy community called jele) to manage the resource and eventually assume more control of fishing by permanent fishers in open waters. In this sense, The NFMP was an ambitious step towards fisheries management as it was, for the first time, the local institutions e.g. village committees were the focus of revitalization while previous *Jalmohal* management discouraged it. This enhanced leasing policy initially seemed to be more effective other management measures but it too finally lead to inequitable measure as well. A number of factors did not favor the process such as the unsuccessful elimination of powerful leaseholders who unofficially control access to many fishing grounds. These former lease holders were in fact the key players of the NFMP process as they took control over processes of access. In many cases fishers are taken out of bidding process by paying lump sums to fisher cooperatives to take control over the fisheries. The only difference between *Jalmohal* policy and NFMP was that previous leaseholders who are mostly outsiders could not bid openly but they are still beneficiaries as they often lend the money to fisher cooperatives to bid on their behalf. The main condition of such lending was that fishers were forced to sell out their catches for lower prices than the market price to the moneylenders/middlemen.

In addition to the "Protection and Conservation of Fish (Amendment) Act (1995), and a more progressive NFMP, there are various national plans and policies that are equally applicable to fish conservation and management of fisheries. For example, Bangladesh National Fifth Five Year Plan (1996-2000) and the National Fish Policy of 1998. These management tools emphasized improved aquatic environmental conditions with respect to biological management and the judicious use of fisheries resources. There are some other pieces of legislation e.g. National Water Resource Planning Act 1992; National Environmental Policy of 1992 and National Water Policy 1997) that are also significant in fisheries development where a number of other agencies are involved such as the Department of Environment (DoE), and Local Government Engineering Department (LGED). Overall, all the related rules, acts, and plans call for effective measures in fisheries areas damaged by other activities such as water resource development and flood control dams and embankments. However, no coordination policies exist between agenceis which can help implement Fish Acts and other related policies.

It is evident that the government of Bangladesh has developed a substantial number of acts, plans and polices relevant to the management of fisheries. However, the related acts, regulatory measures and policies have not effectively improved fisheries management and have not prevented the degradation of fish habitats (SEHD 2003; WorldFish 2006). As such, these policy tools have been highly criticized in various studies for the lack of proper implementation and setting exemplary improvements to inland open water fish catches (Kabir and Khan 2004; Farook 1999). Therefore, there is a need to urgently redefine the fisheries acts, legislation and related policy tools and overhauls the rules and orders if they are to deal with the exiting crisis in the fisheries resources in Bangladesh. Using fishers' knowledge and perception, the following section based on fisher knowledge tries to explain why some important management measures are not very effective in increasing fish production from open waters and ensuring the conservation and restoration of degraded fisheries.

4. Policy Implications of Fisher Knowledge and Observation on Inland Fisheries

It was found during my study that the continuous degradation of open water fisheries and reduction of production have made local fishers and rural farmers concerned about losing their major source of income as well as inexpensive sources of proteins. In addition to high fishing pressure and changes in fish habitats due to man-made causes (e.g. cross dams, lack of migration routes for flood control structures), fishers and other community members mostly blame the ineffectiveness of regulations and management measures under Fish Conservation Act and strategies apply for fisheries development, e.g. banning of certain fishing gears (e.g. gill nets) during the breeding season (May-August). Similarly fishers perceived that several other management measures such as habitat rehabilitation programs are often not effective in the way they are designed. It was also found during the study that fishers are reluctant to comply with the rules and regulations (such as gear banning and reducing fishing efforts) unless they have alternative sources of income and unless they address the real problems for fish decline. The issue of livelihoods is important in the case of effective compliance to fisheries rules (Gezelius 2004) especially for the community fishers who have no alternative sources of income except fishing. Even compliance favored by local norms is only effective if fishers have alternative income except fishing. In the presence of hungry fishers in a community, local supports for enforcement are also absent (Gezelius 2004). This study found that fishers have different opinions in the way the respond to fishing rules. If scrutinized carefully, their observations and claims seem compelling as they are based on the long term observation of the fishers. In the following subsections, I will highlight some of their perspectives with respect to conservation and management of inland fisheries in Bangladesh.

4.1 Fishers' Perception on Habitat Conservation/Restoration Measures

The conservation measures applicable to Bangladesh inland fisheries includes fishing bans through regulation of gears that targets basically on gill-nets in wet season and improving fish habitats by taking measures like restoring of connectivity among habitats and establishing fish sanctuaries (WFC 2006). The motivation behind these types of actions is that fish habitats protection may effectively protect fish resources. As a result, there is increasingly wider establishment of low cost sanctuaries (brush plies or deeper pools at water body managed for fish refuse at dry season) and removing dikes alongside the rivers in order to help fisheries development.

However, not all conservation measures such as developing connectivity are unanimously supported by the fishers. In some cases, the response of fishers with respect to removal of dykes or excavation of degraded canals for habitats rehabilitations is mixed. According to the fishers, a physical connection among habitats ensures timely in- and outflow of waters and improves the fish production. This is very helpful for floodplain preferring fish that spawn early in the monsoon such as snakeheads- shole *Channa striatus*, boal (*Wallago attu*), taki (*Channa puntatus*) and many small indigenous fish such as puti (*Puntius spp.*), tengra (*Mystus spp.*), and baims (*Mastecembellus spp*). In this respect, fishers pointed out that depending on the nature of operation related to the particular measure, such as connectivity there may have negative impacts too. It may for example facilitate an early departure of waters from the feeding ground of fish at recession phase at late monsoon (October-November). Early departure of water takes away juvenile fish from their feeding grounds towards the rivers and thus, reduces the overall production of the *beel* fishery. In that case, a quick increase of catch composed of basically juvenile fish is noticed at late monsoon in some years. This higher catch in late monsoon does not necessarily contribute to a high overall annual production. This happens especially in the years when recession of flood water starts early in September following the early disappearance of monsoon rains. To avoid fish production loss caused by receding water, fishers prefer water controlling structures such as sluice gates in order to maintain adequate water levels during the recession period (October-December).

Like connectivity, fish sanctuaries can play an important role in conserving fish habitats and increase fish production. However, for achieving benefits from such approach, many underlying issues have to be addressed and assessed carefully. Such strategies are yet to be formulated and tested regarding the applicability of types of sanctuaries and maintenance measures to support natural production. There is also lack of understanding about the mode of operation and development of fish sanctuaries and there scaling-up process. At the present, they are operated mainly under project water body or co-managed water body that are small but mostly absent in larger rivers. Both fishers and fisheries managers believe that larger sanctuaries are particularly useful as poaching of fish from such sanctuaries is difficult. However, construction of larger sanctuary is not always a feasible option due to complication related to the bulk acquisition of lands from publicly owned beels (small natural lakes found in floodplain areas). Often fishers refer to the conservation of multiple habitat features (e.g. aquatic vegetations) required for protecting fish during spawning season, as well as in summer when water levels go down. In such cases, fishers pointed out that use of non-agricultural lands like road-side canals and burrow-pits as dry season fish refuges may be an option. To conserve habitats, the majority of fishers support conservation of enough water during dry season. Availability of water in dry seasons may save aquatic vegetations needed for fish production. So far, there are no policies regarding conservation non-traditional fish habitats like borrow-pits.

4.2. Habitat Loss Is Responsible for Fish Decline

In Bangladesh, increased number of fishers is generally cited as the causes for declining of the country's fishery resources (de Graaf 2001; Hughes et al. 1994; Ali 1997; FAP-17 1994). In addition habitat destruction by roads, drainage and flood control, and siltation for upstream deforestation or flow pattern changes, have also contributed to the loss of inland fisheries. Many fishers in the current study area agreed that fish catch have been declining for the above mentioned causes. However, a great majority of fishers hold a different view with respect to the causes of production decline due to increased fishers. Fishers basically emphasize habitat destructions and not the fisher numbers in that respect. The argument related to the habitat loss carries more weight when fishers added the issue of fish biodiversity, migration trends and lack of dry season water refuge that are needed for a healthy production of fish. Fishers informed that due to habitat loss, migration (spawning, feeding, over-wintering) of fish is impaired. As a result, migratory fish (e.g. carps and large shrimp) do not enrich the diversity of catch, and fail to contribute to the fish catch. Often fishers have to rely on several small species of low market value (e.g., small shrimps and minnows). Similarly, dry season habitats do not support broods for a range of deep water preferring species who previously were the regular visitors for breeding and feeding and eventually production of juveniles does not maintain.

In this regard, the fishers' perception is that if the fish habitats are not reduced due to water scarcity and loss of connectivity, fishing pressure by itself cannot damage the resource base - many fish will be able to escape from the gears whatever the number is being operated. The fish saved would be able to propagate in the next season and catch will be guaranteed. In this regard, a detailed study is needed to find out the ways to identify how much habitats could be saved, especially in dry seasons, and how connectivity among habitats could be reestablished (Rahman and Minkin 2003). Fishers deserve more project investment in improving fish habitats.They believe that if fish habitats are properly protected other components of management measures like reducing fishers and implementation of fishing rules would not be of much importance.

It is understandable that the lack of understanding of the issues related to the loss of fish production in natural waters does not help fisheries managers and policy-maker to take appropriate measures in managing fish habitats rather it misleads them to include less effective measures such as prescribing reduction of fishers number what is virtually not possible in country like Bangladesh with limited scope of incomes for many of the rural poor. As such an effective way would be to accommodate fisher knowledge and opinion with respect to policy development.

4.3. Gill Nets Are Not Necessarily Harmful

Policy measures (Conservation of Fish Act 1950 and its amendment 1995) related to the regulation of fishing gears seems to be flawed as per fishers' observation. Bangladesh's inland fisheries are rich in fish diversity which requires a vast diversity of gears if fish are to be harvested effectively. To optimize fishing, a number of traditional techniques have been used historically by the local fishers in open waters (e.g. gillnets, traps and hand operable nests). Under the Fish Act 1950, catching of carps below nine inches long, and catching of juveniles of *hilsha* and several other silurid fish such as local pangash (*Pangusias pangasias)* , silong (*Silonia silondia*)and ayre (*Sperata spp.*), are prohibited during May to September. Such fishing, in general, is carried out using monofilament gill nets, locally called *current jal* for its high efficiency. These nets have been used since the beginning of the 1970s. Since that time, there has been much opposition to the gear. Considering the sizes of catch and damaging effects of the gear, the Government of Bangladesh made a decision to ban gill nets in 2001. Even though there has been support from general public in implementing the net ban, very little progress has been made so far (pers. com Guasuddin Khan, Consultant WorldFish Dhaka,2007). The ban on *current jal* was challenged by the Fisheries Association of Bangladesh which filed an appeal immediately after the decision was made in 2001 that has been held up by the Supreme Court of Bangladesh. As a result, there have been many questions about the effectiveness of the ban.

The present study also gives interesting information regarding gill nets. It was estimated that about 50% of the local stakeholders and 60% of the community fishers believe that fishing with *current jal* is not as harmful as it is thought. They argue that a *current jal* allows fish to grow to a certain size before they are caught except carps and *hilsha.* According to fisher opinion it is not because of the *current jal* that the fish catch is are declining, but because of small meshed seine nets (locally called *kheta ber jal*) that are non-selective and do not allow any fish to escape including juveniles while that is not the case for many types of *current jals* whose mesh size vary from 2-4 cm e.g. koi and puti jal. Fishing using *ber jal* also greatly damages the fish habitats by uprooting aquatic vegetation. In assessing the overall impact of fishing by various *current jal*, it is assumed that the mass decline of fish is perhaps not related to the gill nets but other methods such as dewatering and using small mesh seines. By this argument fishers challenged the existing fishing rules that prohibit the use of *current jal* e.g. Fish Conservation and Protection Act 1950. Today, confiscating *current jal* from fishers and burning them in front of the public have become a common practice by fisheries enforcement teams organized by DoF. It is widely believed that these burning exercises are politically motivated and derogatory to demonstrate general public that government agencies are active and sincere in conserving fish. In doing so, they actually end up punishing the poor fishers who lose an effective and inexpensive fishing method which is often targeting fish of small to medium sizes. Most importantly, the rule related to the banning of *current jal* is exclusively designed to save carp fisheries but truly the carp fishery hardly contribute to the regular catch from open waters any more (See Table 3). According to fishers confiscating or burning *current jal* is not a solution unless their manufacturing is suspended or more damaging nets such as small meshed seine nests (less then 2 mm mesh) are banned as well.

4.4. Decreased Open Water Production a reality

It is widely believed that the vast inland fisheries of Bangladesh have been deteriorating over the 30 years (SEHD 2003). Fish habitat destruction by roads, improvement of drainage for facilitating agriculture, development of flood control structures, and siltation due to deforestation at upper watersheds, along with overfishing, have been commonly cited as the major causes of the deterioration of the country's inland fishery resources (Hughes et al. 1994; Ali 1997). The above changes in the natural habitats have caused different levels of deleterious impacts on inland fisheries. For example, efforts to control floodwaters and drainage improvement and simultaneous uses of wetlands as agricultural lands have resulted in severe loss of water-spaces and fish habitats. Construction of roads across water areas has fragmented many small but important fish habitats such as *beels* (natural depressions of closed and semi-closed types) and separated floodplains from rivers.

The impacts of flood control structures as embankments are very diverse and often not quantifiable. Embankments have both short and long term deleterious impact on inland fisheries. In general, they block the migratory routes of (habitat fragmentation) mature fishes and juveniles respectively to breeding feeding and habitats. Embankments alter the hydrological patterns of interconnected water bodies (e.g. blocking the connectivity between floodplains and rivers) adversely affecting the life cycle of floodplain dependent fishes. Disconnected fish habitats mean less diverse fisheries as migratory fishes can not contribute to the catch. A less diverse fish community is more vulnerable to environmental threats such as disease outbreaks, droughts or late rains and therefore potentially less productive. Other causes of habitat degradation are related to water pollutions from urban areas. For example, discharge of untreated of toxic substances such as textile dyes as well as pollutants from tanneries that have impaired severely the rivers surrounding big cities like Dhaka, Khulna and Chittagong. It is estimated that Bangladesh has lost 45% of its historical fish habitats and the decline of catch rates are 9% per year (SEHD 2003; MPO 1997). Howeever, Fish cacth statistics of Bangladesh shows inceasing trends its around 4% from previous year (FAO-2010). The increment of inland open water fish catch was also not approved by the interviewees in the study area and disagreed by 95% percent of fishers. They believe that fish catch from open water have declined almost 50% in last 30 years. If we consider the habitat loss from different stressors as described above including high fishing pressure in open waters and loss of fish species then projection of an increased fish production from open waters may not be true. The study made by WorldFish Center on 119 open water fisheries areas also have indicated that open water fish catch in uncontrolled water areas are not increasing (WFC 2006). WorldFish found increasing trends in fish catches only from well managed community control waterbodies such as co-managed fisheries (see Rab 2009). Given the loss of habits and considering stressors active on fisheries systems of inland open waters there is a poor chance of increasing production from open waters fisheries and a declining fisheries is more probable (SEHD 2003; WFC 2006). In this regard fishers' observation e.g. loss of open water is perhaps more justified than what is projected in national catch statistics. .

5. The Way Forward: A Shared Vision to Management

We have noticed that fisheries the management of Bangladesh implemented several approaches such as revenue harnessing from fishing, stocking of open waters and developing policies that address habitat degradation problems in open waters and improve socio-economic conditions of fishing dependant people. The inland fisheries resources, however, are widely considered to be in a critical situation with respect to sustainability aspects except some progress in the development of culture-based fisheries in lakes and some close but small-scale fisheries e.g. ponds (Valbo-Jørgensen and Thompson 2007; WFC 2006; Thompson et al. 2000; Mirza and Ericksen 2005). Approaches such as revenue focused *Jalmohal* policy are unsustainable as rent-seeking behaviors the leaseholders that discourage conservation and local institutions. In this regard, there is a need reconsideration of the policy that encourages fisher participation in developing policy. We have also observed that fisheries management and conservation policies are not up to date and are not developed through the consultation of local stakeholders (See section 3). As such fisheries management are top-down, sector-based and mostly ineffective when it deals with issues of open water capture fisheries. As a result, there is a need to revise/redefine the existing policies and strategies related to fisheries development, in general, and saving inland capture fisheries, in particular. A more collaborative process (e.g. comanagement), thus, perhaps a better solution and could be a stating point to initiate dialogue for a better policy development. This may lead to better fisheries management where fishers are the active contributors and management partners with government actors. Given this understanding, the recent management policy would seek alternatives to existing top-down management option that support pragmatic development such as co-management. Co-management approach is more progressive in the sense that it promotes participatory decision making, engages stakeholders from all spheres of fisheries management to hold dialogues on debating issues and ensures a more equitable distribution of resource benefits (WFC 2007). From management point of views, this is more likely a sustaining approach to the governance of vast inland fisheries of Bangladesh. The support for accepting such an approach is, therefore, also triggered by various management failures as described before based on fisheries perceptions. The recent development of common property theory as perceived by various authors working in support local participation also support this multi-stakeholder approach to improve commons (Ostrom 1990; Berkes 1999; Freeman 1992; Johannes et al. 2000). It is understandable from this research that local stakeholders have lots to offer with respect to policy development and implementation of management measures for example fish acts. As a result, the recent management needs are to focus on developing more participatory governance that emphasizes the extended role of local users of common property systems. In this approach, equity and access ensure better livelihoods options and shared decision making facilitate cooperation that may lead to better conservation of resource (Berkes 2006; Valbo-Jørgensen and Thompson 2006; Lobe and Berkes 2004; Pomeroy 1995; Ostrom 1990). This approach also subsumes that communities may promote better protection of the resource, since they live closer to the resource and they have high stakes as fisheries are significant to their livelihoods and part of the culture of many fishing dependant communities such as the *Jele* communities in Bangladesh. Including fishers in management also reinforces the idea of moving from state policing of resources towards an arrangement of joint management that can be a more realistic way of reaching environmental and socioeconomic objectives, such as long term sustainability and maintaining diversity of fish stocks and habitats on which fish production actually depend.

6. Conclusion

Recognizing the existing constraints in recent and past polices applicable to inland open waters of Bangladesh, it is evident that redefining fisheries policy is needed but that it has to be seen within the context broader socio-economic and ecological contexts of the country. There is a higher potential of success if principles of co-management are adapted in the policy process and management interventions and if the fisher livelihoods are considered *a priori*. However, such an approach will only be successful under the conditions if primary stakeholders especially government departments are keen to initiate a multi-stakeholder approach to fisheries management. In this regard, it is essential to develop a dynamic partnership between fishers' local institutions on the one hand and the government institutions on the other to support a greater participatory approach, using the capacities and interest of the former complemented by the ability of the latter to offer enabling legislation and administrative (Hussain et al.2006). Though co-management and its implementation are in its early stages in many countries including Bangladesh, it is a starting point for stakeholders to hold dialogue on fisheries development issues.

Acknowledgments

My special thanks to the community members and other stakeholders of the Fisheries Co-management program of the WorldFish Center, Bangladesh, who gave me the opportunity to discuss various management issues related to fisheries degradation in inland waters. I also would like to express my gratitude to Dr. Fikret Berkes, Canada Research Chair, Natural Resources Institute, University of Manitoba, Winnipeg, Canada, for his assistance in preparing this manuscript. I also equally acknowledge Alan Brooks, Portfolio Director, WorldFish Center, Dr. Paul M. Thomson, Former Project Director, WorldFish Center, and Dr. Parvin Sultana, WorldFish Center, Dhaka, Bangladesh, who extended their support inconducting the research and providing access to the WorldFish Database system and archival records and offered other suggestions that benefited the study. I extend my sincere thanks my friend Ron Jones, FACT Project, Cambodia, for his input in developing and editing the paper. This research was supported by the Canada Research Chair in Community Based Resource Management, held by Dr. Fikret Berkes, Natural Resources Institute, University of Manitoba, Canada (http://www.chairs-chaires.gc.ca), and the WorldFish Center, Dhaka (http://www.worldfishcenter.org).

References

Ahmed, M., Parvin S. and Paul M.T. and C. K. Chong. 2003. *Understanding Livelihoods Dependent on Inland Fisheries in Bangladesh and South East Asia*. WorldFish Center,

Dhaka. [online] URL: *http://www.worldfishcenter.org/Operational%20Plan/OP2003/PDF/op2003_priap.pdf*

Ahmed, N. A. 1999. Fingerling stocking in open waters. *In H. Middendorp, P.Thompson and R. Pomeroy (eds) Sustainable Inland Fisheries Management in Bangladesh. ICLARM Conf. Proc. 58. pp 201-8.*

Ali, M. L. and Paul M. Thompson. *2006.Wetland protection through sanctuaries in Bangladesh.* Draft Technical Paper 4. MACH (Management of Aquatic Ecosystem through Community Husbandry).

Ali, M. Y. 1997. *Fish, Water and people reflections on inland openwater fisheries resources of Bangladesh.* The university press limited, Dhaka, Bangladesh.

Bangladesh Environment Conservation Act, 1995. (Act No. 1 of 1995) (as amended by Act Nos 12 of 2000 and 9 of 2002).

Bangladesh Gazette. 1982. Protection and Conservation (Amendment) Ordinance 1982 (Ordinance No. LV).*An Ordinance to further amend the Protection and Conservation of Fish Act, 1950.* (East Bengal Act XVIII of 1950). Bangladesh Gazette, 15 December 1982, pp. 4991-4993

Bangladesh gazette. 1995. Protection and Conservation of Fish Rules, 1985. *An Ordinance to further amend the Protection and Conservation of Fish Act, 1950.* (East Bengal Act XVIII of 1950). Bangladesh Gazette, 15 December 1982, pp.

BBS (Bangladesh Bureau of Statistics). 2007. *Statistical Pocket Book Bangladesh.* [online] URL*: http://www.sdnpbd.org/sdi/statisticapocketbook/*

Berkes, F. 1986. Local level management and the commons problem: a comparative study of Turkish coastal fisheries. *Marine Policy* 10(3): 215-229

Berkes, F. 1999. *Sacred Ecology: Traditional .Ecological Knowledge and Resource Management.* Philadelphia: Taylor & Francis.

Berkes, F. and C. Folke. 2002. Back to the Future: Ecosystem Dynamics and Local Knowledge. In: *Panarchy: Understanding Transformations in Human and Natural Systems* (Gunderson, L. and C. S. Hollings eds.). Island Press, Washington, DC:121-146.

Berkes, F. 2006. The problematique of community-based conservation in a multi-level world. Paper presented at the Biennial Meeting of the International Association for the Study of Commons (IASC), Bali, Indonesia, June 2006. *http://dlc.dlib.indiana.edu/view/conferences.html.*

Dyer, C. L. & J. R. McGoodwin (eds).1994. *Folk Management in the World's Fisheries: Lessons for Modern Fisheries Management,.* University Press of Colorado, Niwot, USA.

de Graaf, G.J., Born, B. Uddin, K.A. and Marttin, F. 2001. Floods Fish & Fishermen: Eight years of experience with floodplain fisheries, fish migration, fisheries modeling and fish biodiversity in the compartmentalization pilot project, Bangladesh. The University Press Limited, Dhaka, 108 pp.

FAO (Food and Agricultural Organization of UN). 2010. FAO Fisheries and Aquaculture Department Report Bangladesh

FAO (Food and Agricultural Organization of the UN). 1997.Potential constraints and strategies for conservation and management of Inland open water fisheries in Bangladesh. CONSTRAINTS AND STRATEGIES FOR CONSERVATION URL: http://www.fao.org/docrep/x5625e/x5625e07.htm

FAP (Flood Action Plan)-17. 1994. *Flood Fisheries Studies*. Final Report, Main volume. Flood plan coordination organization, Ministry of Water Resource, Government of Bangladesh, Dhaka.

Farook, M. 1999. Analysis of the legal regime for inland fisheries in Bangladesh. *In*:Sustainable Inland fisheries management in Bangladesh. *Proceedings International Center for Living Aquatic Resources Management (ICLARM)* 58:37-43.

Fish Policy Bangladesh. 1998. *Ministry of fisheries and livestocks*, Bangladesh.

Freeman. M.M. R. 1992. The nature and utility of traditional ecological knowledge. *Northern Perspectives* 20(1):9–12.

Gadgil, M. P. Olsson, F. Berkes and C. Folke. 2003. Exploring the role of local ecological knowledge in ecosystem management: Three case studies. In Berkes, F., J. Colding and C. Folke (eds.). *Navigating social-ecological*

Gezelius, Stig (2004) "Food, Money, and Morals: Compliance Among

Natural Resource Harvesters" Human Ecology. Vol 32, no 5, p 615-633.

Hughes, R., S. Adnan and B. Dalal-Clayton. 1994. *Floodplains or Flood Plans? A Review of Approaches to Water Management in Bangladesh*. IIED, London.

Hossain M M., M. A. Islam, S. Ridgway, and T., Matsuishi. 2006. Management of inland open water fisheries resources of Bangladesh: Issues and options. Fisheries Research 77 (2006) 275–284.

Hussain, G. 2010. Freshwater fishes of Bangladesh: fisheries, biodiversity and habitat. *Aquatic Ecosystem Health and Management*, 13: 85-83.

Hviding, E. & Jul-Larsen, E. 1993. *Community-Based Resource Management in Tropical Fisheries*. Centre for Development Studies, University of Bergen, Norway.

IIED (International Institute for Environment and Development). 2010. Local Organizations – An Introduction. http://www.iied.org/governance/key-issues/partnerships/ local-organizations-introduction

Jentoft, S. and T. Kristoffersen. 1989. Fisheries co-management: the case of the Lofoten fishery. *Human Organization* 48 (4):355-365.

Johannes, R.E., M.M.R. Freeman, and R.J. Hamilton. 2000. Ignore fisher's knowledge and miss the boat. *Fish and Fisheries* 1: 257-271

Kabir, M. I. and S. A. Khan. 2004. Management of fish sanctuaries: legal options. The National Holiday. A Bangladesh Daily internet version. [online] URL: *http://www.weekly holiday.net/2004/250604/env.html.*

Keeley, J. 2003. *Policy Processes for Community Based Fisheries Management in Bangladesh: Lessons for Change*. [online] URL: *http://www.ids.ac.uk/ids/bookshop /LESS/BANGLAFISHERIES.pdf*

Lobe, K. and F. Berkes 2004. The padu system of community-based fisheries management: change and local institutional innovation in south India. *Marine Policy* 28:271-281.

Mamun, A. 2006. *A guide to Inland fishes of Bangladesh*. WorldFish Center, Dhaka. 92 pp.

Mamun, A. 2010. Understanding the Value of Local Ecological Knowledge and Practices for Habitat Restoration in Human-Altered Floodplain Systems: A Case from Bangladesh. *Environmental Management*: 922-938

Mamun, A. 2007. *Traditional ecological knowledge and its importance for conservation and management of freshwater fish habitats of Bangladesh.* Masters Thesis. University of Manitoba, Winnipeg, Canada. 100 pp

Minkin, S. F. and J. K. Boyce. 1994. Net losses: "development" drains the fisheries of Bangladesh, *Amicus Journal*, 16 (3):36-40.

Ministry of Water Resources. 1992. Water resource planning act. Published in Bangladesh Gazette Extraordinary Dated 29th January, 1992. URL: *http://www.fao.org/fishery/legalframework/nalo_bangladesh/en*

Mirza, M. Q. and Neil J. Ericksen. 2005. Impact of water control projects on fisheries resources in Bangladesh. *Environmental Management* 20 (4): 437-593

MPO (Master Plan Organization) Bangladesh. 1997. *Openwater capture fisheries resources.* 1987. Technical report 18.pp. 72.

Neilson, J. R., P. Degnbol, K. Viswanathan and M. Ahmed. 2002. Fisheries Co-Management - *An Institutional Innovation: Perspectives and Challenges Ahead.* IIFET 2002 paper no 216

Njaya, F. 2007. Governance Challenges for the Implementation of Fisheries Co-Management: Experiences from Malawi. *International Journal of the Commons Vol 1*, no 1 October 2007, pp. 137-153

Ostrom E. 1990. *Governing the Commons: The Evolution of Institutions for Collective Action.* Cambridge University Press: Cambridge

Parvin, S. and M. F. Islam. 2002. *Open-water Fisheries in Bangladesh: A Critical Review.* [online] URL: *http://www.lars2.org/unedited_papers/unedited_paper/Parveen.pdf*

Pomeroy, R. S. 1995: *Community-based and co-management institutions for sustainable coastal fisheries management in Southeast Asia.* [online] URL:

http://www.co-management.org/download/reprint1.pdf

Rab, M.A. 2009.*River fisheries management in Bangladesh: Drawing lessons from Community Based Fisheries Management (CBFM) experiences*.doi:10.1016/j.ocecoaman.2009.08.001

Rahman, M. and S. F. Minkin. 2003. Net Benefits: The Ecological Restoration of Inland Fisheries in Bangladesh. *Paper presented in international conference of natural resources.* Held in Tagaytay City, the Philippines.

Rahman, A. 1989. *Freshwater Fisheries of Bangladesh.* Bangladesh: Zoological Society of Bangladesh, C/O Department of Zoology, Dhaka University, Dhaka, Bangladesh.

Roos, N., M. M. Islam and S. H. Thilsted. 2003. Small Indigenous Fish Species in Bangladesh: Contribution to Vitamin A, Calcium and Iron Intake. *J. Nutr. 133*(11). [online] URL: *http://jn.nutrition.org/cgi/reprint/133/11/4021S.* [accessed: October 2006].

Ruddle, K. 1994. Changing the focus of coastal fisheries management. In *Community Management and Common Property of Coastal Fisheries in Asia and the Pacific Concepts, Methods and experiences*, ed. R. S. Pomeroy. ICLARM Conference Proceedings 45, Manila, Philippines, pp. 63-86.

SEHD (Society for the Environment and Human Development . 2003. *Bangladesh Environment Facing the 21th Century.* (editor P. Gain, SEHD), Dhaka

Scullion, J. 1996. Flood control, rice and fish in Bangladesh: Some lesson for the lower Mekong Basin. *Mekong fish catch and culture* 2(2): November 1996.

Santha, S. D.(2008)'Local ecological knowledge and fisheries management: a study among riverine fishing communities in Kerala, India', *Local Environment*,13(5): 423 — 435

Smokorwski, K. E., K.J. Withers, and J.R. M. Keslo. 1998. Does habitat creation contribute to management goals ? A evaluation of documented freshwater habitat rehabilitation or enhancement. Canadian technical report of fisheries and aquatic sciences, Department of fisheries ocean. *Greate lakes laboratory of fisheries and aquatic sciences*, ON, Canada.

Thompson, P.M. 2006. Fishers Rights and access: Poverty reduction and sustainable fisheries. MACH project Briefing note 6. Available at: http:*pdf.usaid.gov/pdf_docs/ PNADJ855.pdf*

Thompson, P.M. 2003. Understanding livelihoods dependent on inland fisheries in Bangladesh and Southeast Asia. DFID/FMSP Project R8118. Bangladesh Summary report. [online] URL:*http://www.ahj00.dial.pipex.com/r8118/r8118_13.pdf* [Accessed May 2005].

Thompson, P. M., Sultana, P. and M. N. Islam. 2000. Cooperation, Conflicts and Sustainability in Community-Managed Fisheries in Bangladesh. *In proceedings of the Eighth Conference of the International Association for the Study of Common Property*, Bloomington, Indiana, USA, May 31-June 4.

Thompson, P. M. 1997. Social and distributional issues in open water fisheries management in Bangladesh. *Papers presented at the FAO/DFID Expert Consultation on Inland Fisheries Enhancements Dhaka, Bangladesh*,7–11 April 1997.

Toufique, K. A. 1994. Some observations on power and property rights in inland fisheries of Bangladesh. *World Dev*. 25(3):1-11.

Unhoff, N. 1992. Local institutions and participation for sustainable development. Gatekeeper Series 31. URL: http://pubs.iied.org/pdfs/6045IIED.pdf

Valbo-Jørgensen, J. and Paul M. Thompson. 2007. Culture-based fisheries in Bangladesh: A socio-economic perspective. *FAO Fisheries technical paper*. 499. pp.56

WFC (WorldFish Center, Bangladesh). 2003. *Annual Report on Community-based fisheries management project (Phase-2)*. World Fish Center, Dhaka, Bangladesh, pp. 165-175

WFC (WorldFish Center, Bangladesh). 2006. *Fourth output to purpose review*. Unpublished official report. World Fish Center, Dhaka, Bangladesh, 80 pp.

WFC (WorldFish Center, Bangladesh). 2006. Case Studies of co-managed fisheries. Unpublished official records.150pp.

In: Fishery Management
Editor: Janice S. Intilli
ISBN: 978-1-61209-682-7

Chapter 3

IMPLEMENTING POPULATION VIABILITY ANALYSIS INTO FISHERIES MANAGEMENT

***I. Jarić*[1*], *J. Knežević-Jarić*[2], *G. Cvijanović*[1], *and M. Lenhardt*[3],**
[1]Institute for Multidisciplinary Research, University of Belgrade, Kneza Viseslava 1, Belgrade, Serbia
[2]Ecological Society "Endemit", Oračka 42, Belgrade, Serbia
[3]Institute for Biological Research, University of Belgrade, Despota Stefana 142, Belgrade, Serbia

ABSTRACT

Fish populations worldwide have experienced severe decline, and some of the species are facing a serious threat of extinction. Population simulations are currently the most promising approach for the assessment of extinction risk and the evaluation of alternative management measures. The population viability analysis (PVA) represents a method that is able to cope with uncertainties present in ecological data and to support decision-making process within the field of fishery management. This chapter presents an overview of the development of the PVA within the field of fishery science, along with an evaluation of the success of its application in fishery management, and major challenges in its future implementation. Potential applications of PVA are very diverse and, beside the estimation of sustainable fishery levels, they were used to address a number of different issues, such as the efficiency of the minimum length limits, extinction risk of the target and bycatch species, impact of the climate change, effects of habitat loss and degradation, and other. Despite some unresolved limitations and problems in the PVA development and implementation, scientific knowledge and fishery management have both earned immensely from this process. PVA should be recognized as an important tool for the process of research and management planning and policy development, and as an integral evaluation and planning component within the adaptive management process. If used with necessary caution, PVA can provide invaluable help in current worldwide efforts for achieving sustainable fishery.

* Author to whom correspondence should be addressed: Kneza Viseslava 1, 11000 Belgrade, Serbia, E-mail: ijaric@imsi.rs, phone: +381112078477, fax.:+381113055289.

1. Introduction

Fish are one of the most important natural resources, which provide mankind with a wide spectre of ecosystem goods and services (Olden et al. 2007). However, maximum long-term potential of traditionally exploited marine species (estimated at approximately 80-100 million tonnes/year, excluding cephalopods) has been reached, and the fish production worldwide is estimated to be currently around 85 million tonnes/year (Blanco et al. 2007). The fisheries have expanded their reach worldwide, by fishing in deeper waters and remote areas, as well as by moving on to the previously underexploited fish stocks (Pauly et al. 2002). The majority of fisheries worldwide are fully exploited, overexploited, or recovering (Botsford et al. 1997). According to Reynolds et al. (2005), marine fish have declined by a median 65%, and 29% of fish populations have declined by more than 80%. One of the first impacts of unsustainable fisheries is the removal of large, slower-growing fishes, the reduction of the mean fish total length in an ecosystem, and the change in the ratio of piscivorous to zooplanktivorous fishes, which is a process recognized nowadays as "fishing down marine food webs" (Pauly et al. 2002). Marine ecosystems worldwide are suffering severe depletion of apex predators (Robbins et al. 2006). There is a growing awareness that fish are not only facing severe depletion, but also a threat of extinction through overexploitation (Pauly et al. 2002). In 2008, 37% of the fish species evaluated by IUCN were listed as threatened (Chong et al. 2010).

Furthermore, ineffectiveness of the fisheries management has resulted in a collapse of not only targeted species, but also of hundreds of non-commercial species, which are taken incidentally or through poaching and ghost fishing by lost and abandoned gear (Dayton 1998). Beside the threat through bycatch, non-targeted species may also be threatened by fishing activities that create large disturbances and damages to benthic habitats (Cheung et al. 2005). Large areas of benthic habitat and invertebrate communities have been destroyed by trawling (Dayton 1998). According to Dulvy et al. (2003), the majority of marine extinctions were caused by exploitation (55%) and habitat loss (37%). Inland fisheries are even more vulnerable to collapse, and freshwater fish communities are severely endangered, facing an extinction rate that is much higher than those reported for terrestrial and marine ecosystems (Olden et al. 2007; Olden et al. 2010). Main factors of the decline of both marine and freshwater fish populations are habitat loss and degradation, invasive species, overexploitation, pollution, climate change and the spread of diseases (Butler 2005; Olden et al. 2007; Olden et al. 2010).

Multi-species fisheries usually involve many fishers, gear types, and landing sites over large areas, which make them complex and difficult to assess and monitor and to enforce fisheries regulations (Curtis and Vincent 2008). Common practice is to establish biological reference points, which help managers in their decisions on total allowable catch, and typically determine fishing mortality with reference to some target level of spawning biomass (Tahvonen 2009). They are defined as either maximum or target fishing mortality rates or minimum stock biomass levels (Collie and Gislason 2001). Currently used indices of population condition or viability mainly involve fishing mortality rates, recruitment indices, and biomasses of the stock and of the spawning stock (Rochet 2000). However, some fisheries with complex mixtures of fleets and target species are addressed with comparably complex set of control rules and strategies for estimating the triggers for enforcing these rules (Rice and Legacè 2007). To make a progress toward sustainable fisheries management, there is a need for approaches that can cope with the high species diversity and limited data, which are typical of many fisheries worldwide, especially in tropical regions (Zhou and Griffiths 2008). It is necessary to introduce robust and efficient fisheries assessment frameworks that would integrate biological and socioeconomic information and provide meaningful results under high uncertainty (Eisenack et al. 2006; Curtis and Vincent 2008; De Lara and Martinet 2009).

The main purpose of fish stock predictions is to provide guidance on possible ways of managing fisheries and their supporting ecosystems in a sustainable way (Brander 2003). Nowadays, many fishery management bodies require annual updates of certain kinds of assessments that use particular models throughout assessment process (Brander 2003). However, the models that are commonly applied are often oversimplified, and they frequently focus on a single species modelling and assess simple management measures such as maximum sustainable yield (MSY; Pitcher and Pauly 1998). Fisheries management still relies too heavily on the form of management that originates from the traditional mathematical formalism (Brander 2003), and the extinction risk and socioeconomic data are rarely incorporated explicitly into fisheries assessments (Curtis and Vincent 2008). As stated by Pauly et al. (2002), explicit knowledge on stock status, required by traditional fisheries models, will inherently remain imprecise and error prone. Traditional fisheries models and assessments have played a valuable role in fisheries management, but they should be enhanced by additional or alternative models that would use different time scales, resolution and processes, as well as other scientific disciplines (Brander 2003).

Timely identification of fish species or populations threatened with extinction is urgently needed, so that appropriate management and conservation measures can be defined and implemented (Cheung et al. 2005). Population simulations appear to be currently the most promising approach for assessing extinction risk criteria (Dulvy et al. 2004). The development of population models that are able to assess viability and the extinction risks of a population could be a tool that would bridge the existing gaps in understanding of the fish ecology and life history (Jarić et al. 2010). The population viability analysis (PVA) is a method that is able to cope with uncertainties present in ecological data, and represents an indispensable tool in conservation biology (Boyce 1992; Akçakaya and Sjögren-Gulve 2000). Although it is only infrequently applied to exploited fish species, PVA has a good potential to assess trade-offs in data-poor fisheries and support decision-making (Curtis and Vincent 2008).

This chapter presents an overview of the development of the PVA within the field of fishery science, along with an evaluation of the success of its application in fishery management, and major challenges in its future implementation.

2. Population Viability Analysis

2.1. PVA Definition and Classification

Population viability analysis provides a quantitative modelling and assessment framework that explicitly incorporates stochasticity, uncertainty and extinction risk, and is used to assess the status of a species, predict its responses to management scenarios, and provide directions for the development of species and habitat recovery plans (Curtis and Vincent 2008). PVA uses a population dynamics model and estimates of the model parameters to project the population forward in time, with demographic stochasticity added to the model to provide an indication of the future uncertainty (Wade 2002; Maunder 2004).

There is still a lack of consensus on the definition of PVA. It has been used to designate methods that range from qualitative and verbal processes to mathematically sophisticated simulation models (Reed et al. 2002). However, all methods have in common that they represent some form of examination of the interacting factors that place a population or species at risk (Shaffer 1990). The outcome of PVA model can be an estimation of the decline or the extinction risk, quasi-extinction probability, expected time to extinction, chance of recovery, future population size and an assessment of different impacts of human activities or efficacy of different management scenarios (Boyce 1992; Ludwig 1999; Akçakaya and Sjögren-Gulve 2000; Reed et al. 2002; Reed et al. 2003). Reed et al. (2002) suggested that the use of the term PVA should be narrowed to quantitative modelling.

Dulvy et al. (2004) have reviewed methods that are or may be used for assessing threat status of marine organisms. They make a distinction between the three general extinction risk assessment approaches: 1) Life history and "rule of thumb" approaches, 2) Time-series approaches and 3) demographic approaches based on age- or stage-based schedules of vital rates and fisheries reference points (Dulvy et al. 2004).

Life history and "rule of thumb" (heuristic) approaches based on biological correlates of vulnerability are useful tools for the determination of research and conservation priorities, but they should not be used instead of more formalized assessments when the necessary data are available (Dulvy et al. 2003). The simplest PVAs are deterministic models (Boyce 1992). Dulvy et al. (2004) labelled this group as time-series approaches, and they can be, for instance, models that use deterministic population growth rates to make population projections (Dulvy et al. 2004), or population projections based on catch statistics (Lenhardt et al. 2006). However, such models do not consider the stochasticity of the system and thus can create a bias that is proportional to the degree of intrinsic variance in the system, thus producing strikingly different results than models with incorporated stochasticity (Boyce 1992; Burgman et al. 1988).

Cheung et al. (2005) presented a fuzzy expert system for the development of an index of the intrinsic vulnerability of marine fishes, based on relationships between life history and extinction vulnerability, and validated it by correlations with empirical data. Zhou and Griffiths (2008) described a Sustainability Assessment for Fishing Effects (SAFE) method, which is able to assess the impact of fishing on large numbers of non-target, low-economic value and data-poor species, and for establishing sustainability reference points for management decision support. These methods can be useful for heuristic purposes, or as a preliminary tool, for example in rapid assessments of a large number of populations (Boyce 1992; Dulvy et al. 2004).

Many species extinctions went undetected for a long time after the actual event, which is mostly due to a fact that they were often estimated retrospectively using indirect methods, as well as due to a low direct detection power (Reynolds et al. 2005). According to Dulvy et al. (2003), 80% of marine extinctions were detected by indirect historical comparative methods. Grogan and Boreman (1998) and Jarić et al. (2009*a*) used the sighting frequency methods to assess the probability that a fish population is extinct. This group of methods was introduced by Solow (1993), and later further developed by a number of authors (e.g. Roberts and Solow

2003; McInerny et al. 2006; Jarić and Ebenhard 2010). Although they are not true PVA methods, since they do not estimate the future population viability, only a probability that species is already extinct, they can be useful when the only available data are sighting records (Akçakaya and Sjögren-Gulve 2000).

Commonly applied models in fishery management are yield per recruit models, which require estimates of individual growth rates, natural mortality rates and vulnerability to fishing. These models are, however, generally very simplified. For instance, they do not account for the effect of fishing upon the proportion of mature fish and therefore the reproductive potential (Dulvy et al. 2004). Biological reference points are commonly based on a stock-production or dynamic-pool model, yield-per-recruit analysis, or spawning stock biomass per recruit combined with stock–recruitment data (Collie and Gislason 2001). According to Tahvonen (2009), the commonly used biomass approach is too simple and should be applied only in cases of a lack of population age structure data.

Beside the classic ecological models, the economic fishery models are also frequently used (Tahvonen 2009). The bioeconomic models, such as those by Maroto and Moran (2008), Martinet et al. (2007) and Sarkar (2009), typically assess economic aspects of a fishery and economic drives of overexploitation. Model by Martinet et al. (2007) was based on a viable control theory, where any path that does not respect the established ecological, economic or social constraints, or their combination, was considered to be a crisis state of the fishery.

Spatial models, such as different occupancy models, are used for PVA of metapopulations and, beside the spatial dimension, they can include parameters like dispersal between subpopulations and habitat suitability (Sjögren-Gulve and Hanski 2000). In its simplest form, population dynamics is described by specific state variables defined for each spatial cell, whereas stage-based models can include spatial separation of important life history processes that are linked to different age or size classes (Butler 2005).

Demographically structured models group individuals into distinct classes, either according to their age (age-structured models) or certain stages (stage-structured models), e.g. according to size, weight or developmental state (Akçakaya 2000). Age-structured models are some of the most commonly applied model types in fishery management (e.g. Nieland et al. 2008; De Lara and Martinet 2009; Tahvonen 2009). They can be very useful for the analysis of larger populations, or for populations with simpler social organization, such as fish (Akçakaya 2000).

Bousquet and Le Page (2004) presented a review of the development and use of multi-agent simulations (MAS) for ecosystem management. Multi-agent systems, also called agent-based modelling, are composed of an environment, a set of objects, an assembly of agents that represent the active entities in the system, an assembly of relations that link objects to one another, an assembly of operations and a number of operators (Bousquet and Le Page 2004). They are often used by both ecologists and economists to model integrated ecological and economic components, and also to place more emphasis to the decision-making process of the modelled agents and their social organization.

Individual-based models (IBMs) represent simulations where each individual, its behaviour and fate, are separately modelled (Akçakaya and Sjögren-Gulve 2000). IBMs were developed by ecologists who introduced individuals in a model to understand the role of heterogeneity (Bousquet and Le Page 2004). Due to the complex nature of this approach, they are often performed through computer programming (Lacy 1993), and as simulations became more feasible, use of IBMs became much more frequent (Chambers 1993). IBMs have the flexibility to represent individual differences in age, size, spatial location, and other relevant attributes (Uchmański 1999; Jager et al. 2000), and provide a basis for merging genetics with demography (Chambers 1993). IBMs are the only PVA methods that can successfully model genetic changes in a population, or effects of inbreeding depression (Jager 2001; Jager 2005). IBMs are attractive not only for their heuristic appeal, but also because they are easily parameterized (Butler 2005). These complex models can produce quantities of output that can be difficult to manage, although thoughtful consideration of the necessary output and new visualization tools minimize this problem. Spatially-explicit IBMs are able to incorporate the effect of local environmental features, such as prey abundance, water quality and habitat structure, on marine population dynamics (Butler 2005)

PVA program packages, which have become very popular in the field of conservation biology, are also becoming increasingly applied in the field of the management of exploited and threatened fish populations, and some programs like the RAMAS series and VORTEX are recently being more commonly used (Curtis and Vincent 2008; Gao et al. 2009; Jarić et al. 2010). RAMAS software is able to handle very large population sizes, which is an advantage when it is applied to species with high fecundities, such as fish (Brook et al. 1997). It is also appropriate for cases when there are spatial and habitat suitability factors that should be incorporated in the model (Jarić et al. 2011). On the other hand, RAMAS is not an individual based model, but an age or stage structured model that therefore lacks genetic considerations and more complex demographic patterns. VORTEX is a program package that makes an individual-based Monte Carlo simulation of a population, through the action of deterministic factors and environmental, demographic, and genetic stochasticity (Lacy 1993). It is being regularly used by the World Conservation Union (IUCN) to establish quantitative classification of endangered species (Brook et al. 1997). The MAS models are commonly developed with an object-oriented language, often through software platforms such as StarLogo, NetLogo, Swarm or Ecosim (Bousquet and Le Page 2004). Although there is an ongoing discussion on the precision and reliability of PVA program packages (Taylor 1995; Mills et al. 1996; Reed et al. 2002; Ellner et al. 2002), some studies (e.g. Brook et al. 2000) have found a high level of accuracy of the most commonly used program packages and a high concordance in their results. It is believed that the PVA program packages have a good ability to compare the efficacy of different management scenarios (Ebenhard 2000; Reed et al. 2002). Additional advantage of PVA program packages is their low programming skill requirement.

Bayesian analyses can be considered as a traditional form of PVA (Boyce 1992). As a general rule, they include prior probability distribution of model parameters to establish posterior parameter probability distribution, thus allowing direct incorporation of parameter uncertainty into analysis (Wade 2002). Bayesian analyses provide an efficient framework for including all available information into a single analysis and the representation of uncertainty (Maunder 2004).

Stochastic processes that are relevant to population viability can be categorized into demographic, environmental and genetic stochasticity, and catastrophic events (Lacy 1993). IBMs, demographically structured and spatial models often do have stochasticity included within the model, but not necessarily. Development of models that consider stochasticity of the system is considered crucial in order to increase reliability of the PVA (Boyce 1992).

Recently, development of ecosystem based models is gaining wide popularity. These models are seen as a central type of models applied in fishery assessment in the future

(Pitcher and Pauly 1998). One of the currently best known ecosystem models is Ecopath, a software platform that can be used to compare the trophic structure, trophic pathways, energy flux, and efficiencies among ecosystems or time periods within an ecosystem (Butler 2005). Ecopath is considered to be extremely useful for stock assessment purposes, and can be used to estimate the biomass of groups that cannot be sampled directly. Such biomass values may sometimes be even more reliable than estimates from independent surveys (Pitcher and Pauly 1998). As opposed to these authors, Collie and Gislason (2001) recommended that the use of age-structured multi-species models, which incorporate interactions among a relatively small set of species, will provide greater benefit over ecosystem models, which add realism but generally do not provide readily applicable results for management decision support.

Sensitivity analysis represents a classical step in any simulation methodology (Bousquet and Le Page 2004). It is carried out with alternative values for the parameters that have fixed values (Maunder 2004), in order to identify model parameters which influence the most the performance of the model, and it is usually a part of the assessment of every PVA analysis (Akçakaya and Sjögren-Gulve 2000, Reed et al. 2002). The life history parameters that are the most influential upon the population viability are usually proposed for the highest management and/or research priority (Mills and Lindberg 2002). Sensitivity analysis provides a possibility to compare management options even when there is a lack of knowledge of potentially critical population processes (Curtis and Vincent 2008).

2.2. Application of PVA in Fishery Management

Potential applications of PVA are very diverse and they are often used to address very complex problems (Curtis and Vincent 2008). The most common use is the estimation of sustainable or viable fishery levels (e.g. Tyutyunov et al 2002; Tahvonen 2009; Doukakis et al. 2010). Efficiency of the minimum length limits is also a commonly studied issue (e.g. Kennedy and Sutton 2007; Heppell 2007). With the increasing interest in the establishment of areas closed for fishery, such as marine protected areas, models that assess their efficacy are also gaining popularity (Reynolds et al. 2005). There is a significant number of studies that used models to assess effectiveness of such approach (e.g. Hart 2006; Wielgus et al. 2007).

It is crucial to conduct local or regional assessment of fish extinction risk and causes of threat, in order to institute timely protective management measures (Chong et al. 2010). Allendorf et al. (1997) have made an assessment of the extinction risk for a number of Pacific salmonid stocks, by establishing a following system of ranking thresholds, according to the probability of extinction obtained by PVA:

- A stock estimated by PVA to have a 50% or greater probability of extinction within 5 years would be placed in category A (*very high risk of extinction*);
- A stock with at least a 20% probability of extinction within 20 years would fall into the category B (*high risk of extinction*);
- A stock with at least a 5% probability of extinction within 100 years would fall into the category C (*moderate risk of extinction*).

Scenarios assessing impact of the climate change on aquatic ecosystems are also gaining popularity. For instance, by combining freshwater biodiversity modelling with a hydrological model, Xenopoulos et al. (2005) have assessed the extent of potential freshwater fish species extinctions under climate change scenarios. A number of studies have made an assessment of the impact of fishery on bycatch mortality and extinction risk of non-target species (e.g. De Lara and Martinet 2009; Slooten and Dawson 2010).

A number of studies were dealing with the impact of habitat quality, and the effect of habitat loss and degradation. Butler (2005) developed a spatially-explicit, individual-based model, to address potential effects of ecosystem disturbances and restoration measures on fishery resources. Morita and Yokota (2002) and Jager (2006) developed individual based models to assess the effect of habitat fragmentation on population viability. Marcot et al. (2001) applied Bayesian belief networks to assess population viability of fish species under different land management alternatives.

Bioeconomic models commonly assess influence of market dynamics on stock stability and the risk of "economic extinction". Organization of economic exchanges between resource harvesters and consumers, and the use of economic management tools such as quotas, taxes and permits, are often assessed through MAS type models (Bousquet and Le Page 2004). Curtis and Vincent (2008) combined population viability analysis with economic information to evaluate trade-offs between conservation and long-term income from exploited fish species.

Models that include genetic assessments of fishery management are still less common. Nevertheless, successful examples do exist. Hindar et al. (2006) have made an assessment of the genetic impact of the escaped farmed Atlantic salmon (*Salmo salar*) on the wild salmon populations. Jager (2005) assessed genetic impacts of the long-term stocking measures, while Enberg et al. (2009) analyzed evolutionary change as a response to increased pressure from exploitation. Bearing in mind the time and money constraints of undertaking genetic field experiments, the only feasible way forward in this field will be continued development of computer-based predictive models, which will provide risk assessment across the wide range of scenarios (Hindar et al. 2006).

3. Discussion

Development of PVA in the field of fishery management is still facing a number of unresolved problems. According to Brander (2003), the major problems in making stock assessment models are inherent uncertainty that increases with the time scale of predictions, lack of predictability of environmental variation, variability in growth, recruitment and other population dynamics parameters, as well as the delay in implementation of the results of assessments. Purely biological models often assume oversimplified fishermen's behaviour, while many economic models are biologically oversimplified, without population age structure, which can lead to inaccurate predictions (Tahvonen 2009). Factors that influence variability of fish stocks are too often not being included in traditional models (Brander 2003). Pauly et al. (2002) claimed that it is meaningless to limit fishing mortality by limiting only fishing effort, without considering the catchability coefficient, which expresses the effectiveness of a gear. They have emphasized that it should be an important object of future assessments, since it tends to change through time, as a result of changes in fishing technology. Modelling approaches are also limited by a poor understanding of the population dynamics at small population sizes, in particular the likelihood of the occurence of Allee effects, so they can only estimate quasiextinction or the probability of crossing an arbitrary threshold, e.g. 5% of virgin population size (Dulvy et al. 2004).

One of the key and intrinsic problems of PVA is that it normally deals with complex issues, for which our theoretical and practical knowledge is never complete (Shaffer 1990). The major limitation for successful development of PVA is probably the common lack of suitable and necessary data (Shaffer et al. 2002), especially when they are to be used in detailed models (Boyce 2002). The lack of or poor quality of data for many fish stocks inevitably introduces subjectivity and personal judgment into the analysis (Allendorf et al. 1997). Field studies provide data on life history, habitat related parameters and management implementation, but this is both time and money consuming activity, and only a smaller number of species, usually those charismatic ones or those of significant economic interest, are studied well enough. This is especially true for non-target species and communities, for which fishery scientists and managers often lack the information required to properly assess fishery impacts or to develop adequate management measures (Zhou and Griffiths 2008). In cases when estimates of many demographic rates are unavailable, which is common for most rare and endangered species, a range of values, drawn from more-or-less closely related species, is usually applied (Harwood 2000).

Furthermore, stochasticity and genetics are representing two factors whose simulation is probably posing the biggest problems for PVA practitioners (Boyce 1992; Allendorf and Ryman 2002). According to Pitchford et al. (2007), nonlinearity and stochasticity are unavoidable features of exploited fisheries, and models and management strategies which fail to take both into account ought to be treated with caution. Genetic factors are usually considered too poorly known to be included in the model, and as a result few PVAs incorporate genetic components. Lack of general consensus on criteria that should be used to measure viability, such as the thresholds for acceptable extinction risks and the time-frames applied, is another obstacle for successful implementation of PVA results (Shaffer et al. 2002). This is probably partly caused by the fact that the acceptable level of extinction risk is not so much a scientific decision, as it is a social and practical one (Lande 2002).

Nevertheless, there are some possible strategies which could improve the credibility of a model: providing a rigorous presentation of the structure of the model, comparing results with other types of models, comparing simulated and observed data, and the assessment of the relevance of the model hypothesis (Bousquet and Le Page 2004). It is crucial to establish a consensus on standardized protocols used in the field of PVA. In their review, Schmolke et al. (2010) have presented a list of the elements of the good modelling practice, and the principles of transparent and comprehensive ecological modeling (TRACE) documentation, as a standard format for documenting model development, testing, analysis, and application.

Despite all unresolved limitations and problems in the PVA development and implementation, scientific knowledge and fishery management have both earned immensely from this process. A number of studies have shed light on intrinsic causes of vulnerability of a species. Olden et al. (2007) have found that, in marine habitats, large-bodied marine fish are under greater threat of global extinction, while in the freshwater habitats both small- and large-bodied freshwater species are facing greater risk of extinction than those with the more average sizes. Their results indicated that large-bodied species are primarily threatened by commercial fishing, whereas smaller-bodied fish are threatened mostly by habitat degradation and loss (Olden et al. 2007). Similar results were determined by Dulvy et al. (2003), who found that the fish body size was the most prevalent predictor of vulnerability and higher extinction risk (57%), followed by ecological specialization (28%) and small geographical range and endemism (7%). Large-bodied animals are more sought-after and prone to greater mortality, with lower intrinsic rate of population increase, late maturity, dependence on vulnerable habitat and behaviours that significantly increase catchability (Dulvy et al. 2003; Reynolds et al. 2005; Jarić et al. 2010).

Marine protected areas, as one of the most popular fishery management measures in recent time, will also gain much from the development of PVA models. Field studies for the assessment of their efficiency are too time and money consuming, and often can not provide greater reliability than population models. Wielgus et al. (2007) presented how limited demographic data may be used as an additional tool for the management of marine protected areas. Hart (2006) found that long-term closures, such as marine protected areas, do not lead to an increase of fishery yield in adjacent open fishery area when fishing is at or below the level of MSY. However, they increased yield at high fishery levels, prevented the stock collapse and provided benefit for biodiversity conservation (Hart 2006). Furthermore, Pitchford et al. (2007) have shown, by introducing stochasticity in the previously simple deterministic model, that marine protected areas represent a buffer to the uncertainty in the system, and that they perform better than harvest control rules, which are susceptible to data errors and political interference. Marine reserves are critical because of their value in countering uncertainty and in providing resilience against overexploitation by allowing natural systems to regenerate, as well as due to their role as reservoirs of large, high-valued species (Pitcher and Pauly 1998). According to Reynolds et al. (2005), the establishment of marine protected areas has to be combined with a reduction in overall fishing effort.

There is a conflict between the classic fisheries management rules and extinction risk assessment thresholds, which comes from different types of tolerance to risk between the two disciplines (Rice and Legacè 2007). It was believed that the economic extinction of exploited species (i.e., the point when they are no longer a commercially valuable resource) will occur before the actual extinction, but this is not the case for non-target species caught in multi-species fisheries or those with a high economic value, especially if this value increases as species becomes rare (Dulvy et al. 2003). The discrepancy between the extinction risk and "classic" fishery management assessments also comes from different assessment methods, rules and types of thresholds applied. For instance, in a study by Rice and Legacè (2007), for 87% of examined exploited fish species a serious extinction risk was detected, whereas most of the stocks were still within a zone that allowed fisheries management reference points to indicate that exploitation could continue. It is believed that the true extent of marine extinctions has been underestimated, due to a low detection power (Dulvy et al. 2003). Currently, the global total number of extinct fish species is estimated to be 93 (including 12 extinct in the wild) (Chong et al. 2010).

As stated by Dulvy et al. (2003), in order to avoid listing commercial fish species under international threat criteria, perceived high potential for recovery, high variability and low extinction vulnerability of fish populations have been frequently invoked. Nonetheless, same authors have presented evidence against these paradigms in fish ecology, showing that fish populations are in fact even more vulnerable than terrestrial species, that the influence of their high fecundity and dispersal ability is overestimated, as well as their ability for recovery and the extent of population abundance variability (Dulvy et al. 2003). While the attitude of some decision-makers is that the presence of uncertainty regarding the risk posed to a certain species should lead to precautionary decisions, others argue for the delay of protective measures until there is stronger evidence of the impact of human activities (Slooten and Dawson 2010). However, Dayton (1998) stated that the burden of proof in fisheries management should be reversed, so that those wishing to exploit a resource are the ones that have to demonstrate that it will not cause any ecologically significant long-term changes.

While in the ongoing collapse of various valuable fish stocks some saw failure of mathematical models and quantitative stock assessment methods, this was more related to the failure in regulation of catching activity than in shortcomings of scientific advice (Brander 2003). Optimum effort levels were very rarely implemented in practice (Pauly et al. 2002). Two main reasons for the failure of the approach of maximum sustained yield are the lack of political will to implement it and the imprecise predictions of harvest levels, due to an intrinsic uncertainty of the system (Botsford et al. 1997). According to Ludwig and Walters (2002), maximum sustained yields are an inefficient and rarely achieved goal, which belongs more to the convenient political fiction than to the actual science. It is becoming widely accepted that the optimum harvest levels should be set below the maximum sustained yield levels, in order to reduce the potential risk that harvest becomes a selection pressure towards the maturation at younger age and smaller body size (Kuparinen et al. 2008; Enberg et al. 2009).

Bearing in mind the complexity of the fishery management as a system, with its ecological, economic, social and political aspects, a flexible and reflexive management process seems essential (Brander 2003). Management strategies that are designed explicitly to buffer uncertainty in the system will be able to provide a sustainable and productive fishery (Pitchford et al. 2007). As stated by De Lara and Martinet (2009), whenever one of the ecological, economic or social objectives (i.e., sustainability of the resource base, economic viability and equity in access to the resource) is not met, fisheries face an unsustainable situation. According to Butler (2005), restructuring of marine ecosystem management worldwide will be based on a shift from politically manipulated to science-driven management, and from single-species to ecosystem-based models. On the other hand, Collie and Gislason (2001) claimed that it would be more pragmatic to deliberately simplify multispecies models, to enable easier incorporation of the results into fishery management frameworks.

PVA has a potential in reconciling exploitation with conservation, particularly in data-poor situations (Curtis and Vincent 2008). There is a need to establish a global information system to support data integration and sharing among researchers and conservation practitioners, and to make scientific findings useful to decision-makers (Olden et al. 2010). According to Butler (2005), the main job of model developers is to better understand the ecological processes that limit fish stocks.

Dulvy et al. (2003) have recommended that the fisheries scientists should be more proactive in the area of vulnerability prediction, extinction risk assessment and species prioritization for particular conservation or management attention. Dulvy et al. (2004) suggested a two-tiered approach for extinction risk assessment, where the first step would be to use simple methods to rapidly assess large numbers of populations and species, thus identifying those that are potentially vulnerable, and then to subject those identified as vulnerable to more detailed and rigorous population viability analysis. Since it is estimated that only a few marine organisms (<5% of fish species) have been assessed using threat criteria (Dulvy et al. 2004), future PVA development should be extended to cover a much larger number of species. According to Xenopoulos et al. (2005), ongoing climate change and its impact on marine and freshwater fish stocks should represent another future PVA priority.

4. Conclusion

Due to an expressed need for urgent management and conservation measures for declining or collapsed fish stocks, the development of PVA models should not wait for better data. On the contrary, PVA can provide directions to investigators regarding the research activities of greatest importance (Jarić et al. 2009*b*), and to compare the efficacy of different management scenarios (Ebenhard 2000; Reed et al. 2002). Adaptation of a model to reality is generally not performed to convert it into a prediction tool, but rather to understand the dynamics that exist or have existed, and in that way to serve as a decision support tool (Bousquet and Le Page 2004). Therefore, PVA should be recognized as an important tool for the process of research and fisheries management planning and policy development, and as an integral evaluation and planning component within the adaptive management process. Figure 1 presents a general outline of this process. New insights acquired through research and monitoring could be used to evaluate and refine the models, as well as to define further modelling directions. A productive interaction among the PVA developers, scientific community, management authorities, policy developers and the general public will be essential for the general effectiveness of this approach (Jarić et al. 2011). If used with necessary caution, PVA can provide invaluable help in current worldwide efforts for achieving sustainable fishery.

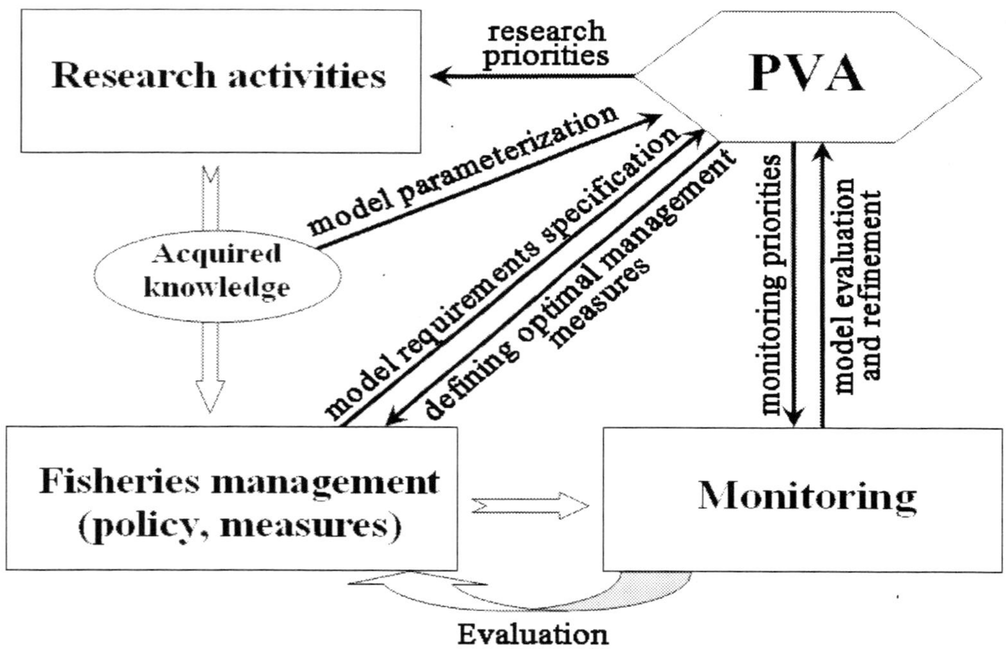

Figure 1. Principles of the population viability analysis (PVA) integration within the research and management planning and policy development processes.

5. ACKNOWLEDGMENTS

We acknowledge the support by Project No. 173045, funded by the Ministry of Education and Science of Republic Serbia.

REFERENCES

Akçakaya, H. R. (2000). Population viability analyses with demographically and spatially structured models. *Ecological Bulletins 48*, 23-38.

Akçakaya, H. R. and Sjögren-Gulve, P. (2000). Population viability analyses in conservation planning: an overview. *Ecological Bulletins 49*, 9-21.

Allendorf, F. W. and Ryman, N. (2002). The role of genetics in population viability analysis. In R. Beissinger and D. R. McCullough (Eds.), *Population Viability Analysis*, 50-85. Chicago and London: The University of Chicago Press.

Allendorf, F. W., Bayles, D., Bottom, D. L., Currens, K. P., Frissell, C. A., Hankin, D., Lichatowich, J. A., Nehlsen, W., Trotter, P. C. and Williams, T. H. (1997). Prioritizing Pacific salmon stocks for conservation. *Conservation Biology 11(1)*, 140-152.

Blanco, M., Sotelo, C. G., Chapela, M. J. and Pérez-Martín, R. I. (2007). Towards sustainable and efficient use of fishery resources: present and future trends. *Trends in Food Science and Technology 18*, 29-36.

Botsford, L. W., Castilla, J. C. and Peterson, C. H. (1997). The management of fisheries and marine ecosystems. *Science 277*, 509-515.

Bousquet, F. and Le Page, C. (2004). Multi-agent simulations and ecosystem management: a review. *Ecological Modelling 176*, 313-332.

Boyce, M. S. (1992). Population viability analysis. *Annual Review of Ecology and Systematics 23*, 481-506.

Boyce, M. S. (2002). Reconciling the small-population and declining-population paradigms. In R. Beissinger and D. R. McCullough (Eds.), *Population Viability Analysis*, 41-49. Chicago and London: The University of Chicago Press.

Brander, K. (2003). What kinds of fish stock predictions do we need and what kinds of information will help us to make better predictions? *Scientia Marina 67(Suppl. 1)*, 21-33.

Brook, B. W., Lim, L., Harden, R. and Frankham, R. (1997). Does population viability analysis software predict the behaviour of real populations? A retrospective study on the Lord Howe Island woodhen *Tricholimnas sylvestris* (Sclater). *Biological Conservation 82*, 119-128.

Brook, B. W., O'Grady, J. J., Chapman, A. P., Burgman, M. A., Akçakaya, H. R. and Frankham, R. (2000). Predictive accuracy of population viability analysis in conservation biology. *Nature 404*, 385-387.

Burgman, M. A., Akçakaya, H. R. and Loew, S. S. (1988). The use of extinction models for species conservation. *Biological Conservation 43*, 9-25.

Butler IV, M. J. (2005). Benthic fisheries ecology in a changing environment: unraveling process to achieve prediction. *Aquatic Living Resources 18*, 301-311.

Chambers, R. C. (1993). Phenotypic variability in fish populations and its representation in individual-based models. *Transactions of the American Fisheries Society 122*, 404-414.

Cheung, W. W. L., Pitcher, T. J. and Pauly, D. (2005). A fuzzy logic expert system to estimate intrinsic extinction vulnerabilities of marine fishes to fishing. *Biological Conservation 124*, 97-111.

Chong, V. C., Lee, P. K. Y. and Lau, C. M. (2010). Diversity, extinction risk and conservation of Malaysian fishes. *Journal of Fish Biology 76*, 2009-2066.

Collie, J. S. and Gislason, H. (2001). Biological reference points for fish stocks in a multispecies context. *Canadian Journal of Fisheries and Aquatic Sciences 58*, 2167-2176.

Curtis, J. M. R. and Vincent, A. C. J. (2008). Use of population viability analysis to evaluate CITES trade-management options for threatened marine fishes. *Conservation Biology 22(5)*, 1225-1232.

Dayton, P. K. (1998). Reversal of the burden of proof in fisheries management. *Science 279(5532)*, 821-822.

De Lara, M. and Martinet, V. (2009). Multi-criteria dynamic decision under uncertainty: a stochastic viability analysis and an application to sustainable fishery management. *Mathematical Biosciences 217*, 118-124.

Doukakis, P., Babcock, E. A., Pikitch, E. K, Sharov, A. R, Baimukhanov, M., Erbulekov, S., Bokova, Y. and Nimatov, A. (2010). Management and recovery options for Ural River beluga sturgeon. *Conservation Biology 24(3)*, 769-777.

Dulvy, N. K., Sadovy, Y. and Reynolds, J. D. (2003). Extinction vulnerability in marine populations. *Fish and Fisheries 4*, 25-64.

Dulvy, N. K., Ellis, J. R., Goodwin, N. B., Grant, A., Reynolds, J. D. and Jennings, S. (2004). Methods of assessing extinction risk in marine fishes. *Fish and Fisheries 5*, 255–276.

Ebenhard, T. (2000). Population viability analyses in endangered species management: the wolf, otter and peregrine falcon in Sweden. *Ecological Bulletins 48*, 143-163.

Eisenack, K., Scheffran, J. and Kropp, J. P. (2006). Viability analysis of management frameworks for fisheries. *Environmental Modeling and Assessment 11*, 69-79.

Ellner, S. P., Fieberg, J., Ludwig, D. and Wilcox, C. (2002). Precision of population viability analysis. *Conservation Biology 16(1)*, 258-261.

Enberg, K., Jørgensen, C., Dunlop, E. S., Heino, M. and Dieckmann, U. (2009). Implications of fisheries-induced evolution for stock rebuilding and reccovery. *Evolutionary Applications 2*, 394-414.

Gao, X., Brosse, S., Chen, Y., Lek, S. and Chang, J. (2009). Effects of damming on population sustainability of Chinese sturgeon, *Acipenser sinensis*: evaluation of optimal conservation measures. *Environmental Biology of Fishes 86*, 325-336.

Grogan, C. S. and Boreman, J. (1998). Estimating the Probability That Historical Populations of Fish Species Are Extirpated. *North American Journal of Fisheries Management 18*, 522–529.

Hart, D. R. (2006). When do marine reserves increase fishery yield? *Canadian Journal of Fisheries and Aquatic Sciences 63*, 1445-1449.

Harwood, J. (2000). Risk assessment and decision analysis in conservation. *Biological Conservation 95*, 219-226.

Heppell, S. S. (2007). Elasticity analysis of green sturgeon life history. *Environmental Biology of Fishes 79*, 357–368.

Hindar, K., Fleming, I. A., McGinnity, P. and Diserud, O. (2006). Genetic and ecological effects of salmon farming on wild salmon: modelling from experimental results. *ICES Journal of Marine Science 63*, 1234-1247.

Jager, H. I. (2001). Individual variation in life history characteristics can influence extinction risk. *Ecological Modelling 144*, 61–76.

Jager, H. I. (2005). Genetic and demographic implications of aquaculture in white sturgeon (*Acipenser transmontanus*) conservation. *Canadian Journal of Fisheries and Aquatic Sciences 62*, 1733-1745.

Jager, H. I. (2006). Chutes and ladders and other games we play with rivers. I. Simulated effects of upstream passage on white sturgeon. *Canadian Journal of Fisheries and Aquatic Sciences 63*, 165-175.

Jager, H. I., Lepla, K., Chandler, J., Bates, P. and Van Winkle, W. (2000). Population viability analysis of white sturgeon and other riverine fishes. *Environmental Science and Policy 3*, S483-S489.

Jarić, I. and Ebenhard, T. (2010). A method for inferring extinction based on sighting records that change in frequency over time. *Wildlife Biology 3* (in press) doi:10.2981/09-044

Jarić, I., Lenhardt, M., Cvijanović, G. and Ebenhard, T. (2009*a*). *Acipenser sturio* and *Acipenser nudiventris* in the Danube – extant or extinct? *Journal of Applied Ichthyology 25*, 137-141.

Jarić, I., Lenhardt, M., Cvijanović, G. and Ebenhard, T. (2009*b*). Population viability analysis and potential of its application to Danube sturgeons. *Archives of Biological Sciences 61(1)*, 123-128.

Jarić, I., Ebenhard, T. and Lenhardt, M. (2010). Population Viability Analysis of the Danube sturgeon populations in a VORTEX simulation model. *Reviews in Fish Biology and Fisheries 20(2)*, 219-237.

Jarić, I., Knežević-Jarić, J., Cvijanović, G. and Lenhardt, M. (2011). Population viability analysis of the European Atlantic sturgeon (*Acipenser sturio* L.) from the Gironde Estuary system. In P. Williot et al. (Eds.), *Biology and Conservation of the European sturgeon Acipenser sturio L, 1758*. Springer-Verlag Berlin Heidelberg. doi:10.1007/978-3-642-20611-5_46 (in press)

Kennedy, A. J. and Sutton, T. M. (2007). Effects of harvest and length limits on shovelnose sturgeon in the upper Wabash River, Indiana. *Journal of Applied Ichthyology 23*, 465-475.

Kuparinen, A., O'Hara, R. B. and Merilä, J. (2008). The role of growth history in determining age and size at maturation in exploited fish populations. *Fish and Fisheries 9*, 201-207.

Lacy, R. C. (1993). VORTEX: a computer simulation model for population viability analysis. *Wildlife Research 20*, 45-65.

Lande, R. (2002). Incorporating stochasticity in population viability analysis. In R. Beissinger and D. R. McCullough (Eds.), *Population Viability Analysis*, 18-40. Chicago and London: The University of Chicago Press.

Lenhardt, M., Jaric, I., Kalauzi, A. and Cvijanovic, G. (2006). Assessment of extinction risk and reasons for decline in sturgeon. *Biodiversity and Conservation 15*, 1967-1976.

Ludwig, D. (1999). Is it meaningful to estimate a probability of extinction? *Ecology 80(1)*, 298-310.

Ludwig, D. and Walters, C. J. (2002). Fitting population viability analysis into adaptive management. In R. Beissinger and D. R. McCullough (Eds.), *Population Viability Analysis*, 511-520. Chicago and London: The University of Chicago Press.

Marcot, B. G., Holthausen, R. S., Raphael, M. G., Rowland, M. M. and Wisdom, M. J. (2001). Using Bayesian belief networks to evaluate fish and wildlife population viability under land management alternatives from an environmental impact statement. *Forest Ecology and Management 153*, 29-42.

Maroto, J. M. and Moran, M. (2008). Increasing marginal returns and the danger of collapse of commercially valuable fish stocks. *Ecological Economics 68*, 422-428.

Martinet, V., Thébaud, O. and Doyen, L. (2007). Defining viable recovery paths toward sustainable fisheries. *Ecological Economics 64*, 411-422.

Maunder, M. N. (2004). Population viability analysis based on combining Bayesian, integrated, and hierarchical analyses. *Acta Oecologica 26*, 85-94.

McInerny, G. J., Roberts, D. L., Davy, A. J. and Cribb, P. J. (2006). Significance of sighting rate in inferring extinction and threat. *Conservation Biology 20(2)*, 562-567.

Mills, L. S. and Lindberg, M. S. (2002). Sensitivity analysis to evaluate the consequences of conservation actions. In R. Beissinger and D. R. McCullough (Eds.), *Population Viability Analysis*, 338-366. Chicago and London: The University of Chicago Press.

Mills, L. S., Hayes, S. G., Baldwin, C., Wisdom, M. J., Citta, J., Mattson, D. J. and Murphy, K. (1996). Factors leading to different viability predictions for a grizzly bear data set. *Conservation Biology 10(3)*, 863-873.

Morita, K. and Yokota, A. (2002). Population viability of stream-resident salmonids after habitat fragmentation: a case study with white-spotted charr (*Salvelinus leucomaenis*) by an individual based model. *Ecological Modelling 155*, 85-94.

Nieland, J. L., Hansen, M. J., Seider, M. J. and Deroba, J. J. (2008). Modeling the sustainability of lake trout fisheries in eastern Wisconsin waters of Lake Superior. *Fisheries Research 94*, 304-314.

Olden, J. D., Hogan, Z. S. and Zanden, M. J. V. (2007). Small fish, big fish, red fish, blue fish: size-biased extinction risk of the world's freshwater and marine fishes. *Global Ecology and Biogeography 16*, 694-701.

Olden, J. D., Kennard, M. J., Leprieur, F., Tedesco, P. A., Winemiller, K. O. and García-Berthou, E. (2010). Conservation biogeography of freshwater fishes: recent progress and future challenges. *Diversity and Distributions 16*, 496-513.

Pauly, D., Christensen, V., Guénette, S., Pitcher, T. J., Sumaila, U. R., Walters, C. J., Watson, R. and Zeller, D. (2002). Towards sustainability in world fisheries. *Nature 418*, 689-695.

Pitcher, T. J. and Pauly, D. (1998). Rebuilding ecosystems, not sustainability, as the proper goal of fishery management. In T. Pitcher, D. Pauly and P. Hart (Eds.), *Reinventing Fisheries Management*, 311-325. London: Kluwer Academic, Fish and Fisheries Series.

Pitchford, J. W., Codling, E. A. and Psarra, D. (2007). Uncertainty and sustainability in fisheries and the benefit of marine protected areas. *Ecological Modelling 207*, 286-292.

Reed, D. H., O'Grady, J. J., Brook, B. W., Ballou, J. D. and Frankham, R. (2003). Estimates of minimum viable population sizes for vertebrates and factors influencing those estimates. *Biological Conservation 113*, 23-34.

Reed, J. M., Mills, L. S., Dunning Jr., J. B., Menges, E. S., McKelvey, K. S., Frye, R., Beissinger, S. R., Anstett, M. -C. and Miller, P. (2002). Emerging issues in Population Viability Analysis. *Conservation Biology 16(1)*, 7-19.

Reynolds, J. D., Dulvy, N. K., Goodwin, N. B. and Hutchings, J. A. (2005). Biology of extinction risk in marine fishes. *Proceedings of the Royal Society B: Biological Sciences 272*, 2337-2344.

Rice, J. C. and Legacè, È. (2007). When control rules collide: a comparison of fisheries management reference points and IUCN criteria for assessing risk of extinction. *ICES Journal of Marine Science 64*, 718-722.

Robbins, W. D., Hisano, M., Connolly, S. R. and Choat, J. H. (2006). Ongoing collapse of coral-reef shark populations. *Current Biology 16*, 2314-2319.

Roberts, D. L. and Solow, A. R. (2003). When did the dodo become extinct? *Nature 426*, 245.

Rochet, M. –J. (2000). May life history traits be used as indices of population viability? *Journal of Sea Research 44(1-2)*, 145-157.

Sarkar, S. (2009). Optimal fishery harvesting rules under uncertainty. *Resource and Energy Economics 31*, 272-286.

Schmolke, A., Thorbek, P., DeAngelis, D. L. and Grimm, V. (2010). Ecological models supporting environmental decision making: a strategy for the future. *Trends in Ecology and Evolution 25(8)*, 479-486.

Shaffer, M. L. (1990). Population Viability Analysis. *Conservation Biology 4(1)*, 39-40.

Shaffer, M., Watchman, L. H., Snape III, W. J. and Latchis, I. K. (2002). Population viability analysis and conservation policy. In R. Beissinger and D. R. McCullough (Eds.), *Population Viability Analysis*, 123-142. Chicago and London: The University of Chicago Press.

Sjögren-Gulve, P. and Hanski, I. (2000). Metapopulation viability analysis using occupancy models. *Ecological Bulletins 48*, 53-71.

Slooten, E. and Dawson, S. M. (2010). Assessing the effectiveness of conservation management decisions: likely effects of new protection measures for Hector's dolphin (*Cephalorhynchus hectori*). *Aquatic Conservation: Marine and Freshwater Ecosystems 20*, 334-347.

Solow, A. R. (1993). Inferring extinction from sighting data. *Ecology 74*, 962–964.

Tahvonen, O. (2009). Economics of harvesting age-structured fish populations. *Journal of Environmental Economics and Management 58*, 281-299.

Taylor, B. L. (1995). The reliability of using population viability analysis for risk classification of species. *Conservation Biology 9(3)*, 551-558.

Tyutyunov, Y., Senina, I., Jost, C. and Arditi, R. (2002). Risk assessment of the harvested pike-perch population of the Azov Sea. *Ecological Modelling 149*, 297-311.

Uchmański, J. (1999). What promotes persistence of a single population: an individual-based model. *Ecological Modelling 115*, 227-241.

Wade, P. R. (2002). Bayesian population viability analysis. In R. Beissinger and D. R. McCullough (Eds.), *Population Viability Analysis*, 213-238. Chicago and London: The University of Chicago Press.

Wielgus, J., Ballantyne IV, F., Sala, E. and Gerber, L. R. (2007). Viability analysis of reef fish populations based on limited demographic information. *Conservation Biology 21(2)*, 447-454.

Zhou, S. and Griffiths, S. P. (2008). Sustainability Assessment for Fishing Effects (SAFE): a new quantitative ecological risk assessment method and its application to elasmobranch bycatch in an Australian trawl fishery. *Fisheries Research 91*, 56-68.

Xenopoulos, M. A., Lodge, D. M., Alcamo, J., Märker, M., Schulze, K. and Van Vuuren, D. P. (2005). Scenarios of freshwater fish extinctions from climate change and water withdrawal. *Global Change Biology 11*, 1557-1564.

In: Fishery Management
Editor: Janice S. Intilli

ISBN: 978-1-61209-682-7

Chapter 4

TREE-BASED MACHINE LEARNING ANALYSIS FOR FISHERIES RESEARCH

Manuel Mendoza, Maria Grazia Pennino and Jose Maria Bellido

Instituto Español de Oceanografía (IEO), Centro Oceanográfico de Murcia, C/ Varadero 1, San Pedro del Pinatar (Murcia), Spain

ABSTRACT

We show how to apply Classification and Regression Trees to deal with problems in fisheries research. Classification and Regression Trees are techniques of Machine Learning that allow the recognition of complex patterns in data. This permits the extraction of hidden information contained in large databases, which represents a major challenge in fisheries research. We illustrate our explanation with a non-published case of study as well as leaning on another example already published by the authors, although very different kinds of study can be treated with these methods. The application process is made up of three main stages. The result of a first data pre-processing phase is the training data set. The second stage involves the application of one or several induction algorithms, which translates into a set of trees. The analysis and interpretation of these trees, in the third and last stage, allows the generation of new knowledge about the system analyzed.

INTRODUCTION

Classification and Regression Trees (CART) are two related techniques of Machine Learning, a scientific discipline that is concerned with the design and development of algorithms that allow computers to learn to recognize complex patterns based on data (Brieman et al., 1993). Although there are some new tree-based methods based on ensembles or networks of trees, like TreeNet and Random Forest, the most commonly used tree-based methods, and the ones that we are going to deal with here, are Classification and Regression Trees.

In machine learning, an "expert system" learns inductively from the input of a series of samples or training data. Unlike other methods considered as "black box" techniques, such as neuronal networks, CART are considered "white box" or "knowledge discovery" techniques because the product of the learning can be represented explicitly (Srinivasu *et al.* 2009; James and James, 2010). Thus, CART can be used to make predictions, and also to extract some knowledge about the system analyzed. They are data mining or data-exploring tools.

CART originated in the statistics discipline, although its first implementation, known as C4.5, was independently developed by artificial intelligence researchers (Quinlan, 1985). Thus, although CART are mainly tools of machine learning, they are still based on statistics. Given that CART applications originated as methodologies of machine learning, their terminology evolved from that of the artificial intelligence. However, given that they can be used to handle the same problems than the multivariate statistics, statistics terminology is also suitable. The present article mixes both terminologies, which make it more understandable for who are more familiar to statistics terms.

Classification or regression trees have to be used depending on the nature of the dependent variable as categorical or numerical respectively. Independent variables can also be either categorical or numerical. Problems commonly dealt with discriminant analysis can be also handled with classification trees and those dealt with multiple or logistic regression can be handled as well with regression trees. The approach, however, is different. These multivariate statistics techniques generate equations (linear combination of the independent variables), which very often are not easy to interpret. CART, however, provide branching trees, sets of if-then logical conditions (splits), which interpretation is straight forward (Mendoza 2007, Rogan et al. 2008). These tree models classify cases into groups or predict values of a dependent variable (criterion), based on values or categories of the independent variables (predictors). The procedure provides validation tools for classification and exploratory analysis or data mining.

Some of the advantages of CART are that they look at variables hierarchically rather than simultaneously, both numerical and categorical data can be used. CART are also easy to understand and interpret, and they do not assume that the dependent variable follows any given distribution. While statistic techniques allow to asses if a given independent variable affects or not on the dependent variable, CART even allow to identify in what conditions of the rest of variables affects. Instead of the effect of the variables, CART allows to recognize the effect of different combinations of variables. Tree models, in addition, can be validated using statistical tests, which makes it possible to account for the reliability of the model. These modelling techniques, moreover, are flexible enough to handle complex problems with multiple interacting elements and typically outcompete traditional approaches (e.g., generalized linear models), making them ideal for fisheries research.

Potential applications of CART for fisheries research are huge. Pons et al (2009) apply different Tree-Based Machine Learning techniques to predict the bycath of loggerheads turtles, from geographical and technical characteristics and environmental conditions of fishing; Wang (2005) applies CART to determine the habitat associations of three commercially important and currently overexploited species, yelloweye rockfish, canary rockfish, and lingcod; Simmonds et al. (1996) uses similar techniques of machine learning for species-recognition from data on the acoustic backscattering coefficients, characteristic of aggregations of mackerel, horse mackerel, and two sizes of cod; Fernandes (2009) applies classification trees for species identification of fish-school to the multifrequency information from the ground-truthed echotraces; Olden et al. (2008) apply classification trees to identify what combination of life history and ecology traits characterize rarity and extinction risk; and the number of applications of CART and other techniques of machine learning to deal with fisheries research and management increases each year.

The aim of this study is to explain how to apply CART, illustrating our explanation with two studies of fisheries research. One of these studies has been recently published by the authors (Mendoza *et al.*, 2010) and the other one is a new non published study. In Mendoza *et al.* (2010), classification trees were used to analyse the effect of closed seasons on posterior European hake yields. The new case of study consists in the application of regression trees to establish those factors that determine the yield of Surmullet (*Mullus surmuletus Linnaeus*, 1758) in the western Mediterranean Sea.

CART APPLICATION

The application process is made up of three main stages (figure 1). A massive data set has to be previously collected with the intention of containing the information or knowledge that we want to extract

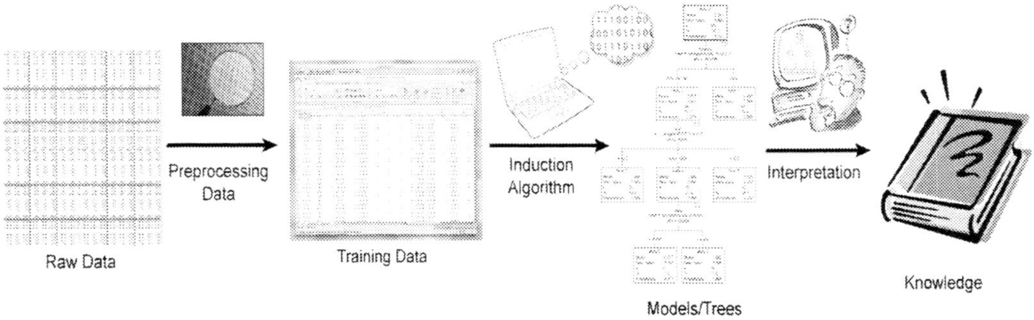

Figure 1. CART application process.

Frequently, raw data do not allow extracting the information that we are looking for, and some kind of pre-processing of data is required. The result of a first data pre-processing phase is the training data set. The second stage involves the application of one or several induction algorithms, which translates into a set of trees. The analysis and interpretation of these trees, in the third and last stage, allows the generation of new knowledge about the system analyzed.

Data Pre-Processing

Data cleaning. The representation and quality of the instance data is essential for the success of machine learning on a given task. Moreover, knowledge discovery during the training phase is more difficult if there is irrelevant, noisy or unreliable information. Identifying and removing irrelevant information, selecting a subset of features, reduces the dimensionality of the data which often allows learning algorithms operate more effectively. In our case of study, the data sets used were extracted from fishing observer data included in the SIRENO (Seguimiento Integrado de los Recursos Naturales Oceánicos) database, developed by the Spanish Institute of Oceanography. This raw data set includes the abundance of a number of species caught in the western Mediterranean Sea since 2003 until 2008, and many features related with the fishery units, the vessels, the environmental and geographic information of the fishing trips, etc. For the study, part of this information was selected and classified by hauls. For each haul we selected the catch of the target species, the Surmullet (*Mullus surmuletus*), and those features that could be related with the variability of its catch. The result was a data set, made up by a number of hauls with their basic features, such as the date, the geographical coordinates and the depth of the starting and ending points. It also includes some environmental conditions, type of fishing gear, and the catch of the target species.

Variables construction. Like in using traditional multivariate statistics techniques, the training data set has to include several independent variables and one dependent variable. The basic features set, however, very often cannot be directly used as variables for the analyses, but these variables have to be constructed from them. For example, in our case of study, the depth of both starting and ending points were not directly used as variables. They were used to obtain the mean depth of the haul, used indeed as independent variable. Instead of the geographical coordinates of the starting and ending points, the final set includes the coordinates of the mean point of each haul. From the timing of the starting and ending points, we computed the hour of the mean point, and the duration of the haul, which was used to standardize the catch, obtaining in that way the dependent variable CPUE (Catch Per Unit Effort, kg/h). The resulting data set was made up by 9 variables: day, month and season of the year, longitude, latitude, depth and hour of the mean point of the haul, wind force (Beaufort scale), and period of the day (early morning: 4-8; morning: 8-12; afternoon: 12-16).

Sometimes, data need a more radical transformation before to be used. The original data in Mendoza *et al.* (2010), for example, were daily landings of hake from 2001 to 2008 in nine landing ports in the south of Spain, together with their respective months of inactivity. The aim of the study was to assess the effect of closed seasons on posterior European hake yields, and CART were the methodology with which to carry it out. For each fish market, mean yields were calculated for every period between closed seasons from 2001 to 2008, but these values of yield were not comparable because the fishing fleet of each landing port and their fishing grounds are different. To solve this problem, the mean annual yield of each fish market was divided by the mean yield of the entire study period for this particular fish market. In this way, a relative yield was obtained for each year for each fish market, which was indeed comparable, not only with the rest of the years in this fish market, but also with the relative yields of the other fish markets. This relative yield, which shows how "good" this particular year was in relation to the rest of the years, in this specific fish market, constituted the dependent variable for the CART analyses. Whether fishing was closed or not in each month constituted the main independent variables (see Mendoza *et al.*, 2010).

Applying methodologies that use all the information supplied by the training set, such as the neural networks, redundant information hinders the process of learning. However, for methodologies like the classification or regression trees, which are able to select those variables involved in the model, redundant information is not a problem. For example, in our original data sets, the date is available for each haul, from which it is possible to construct at least 3 independent variables; the day of the year (from 1 to 365), the month, and the season. The three variables were included in the training set, despite the month and the season can be obtained from the day of the year. If the season is the determinant factor of the abundance of the species, for example, the variable season will be statistically more significant than the other two, and the only one included in the resulting models. In absentia of the season, the month could be finally involved by some tree, but it is also possible that the month variable is not significant enough to be included, mainly if the limits of the seasons does not match with the limit of some of the intervals. Thus, it is worth to include independent variables with similar information, when these supply some additional information regard each other.

Outliers detection. It is also an important step for data pre-processing. Outliers are those values that seem to differ by a substantial amount from the rest of the data. These are often points worthy of investigation in order to understand why they differ. Such points can lead to significant discoveries, although very often they are simple mistakes. When the number of samples is not very high, outliers can distort the results, so it is important to detect the outliers for being analyzed before to use the final training set. One way for detecting outliers is representing different combinations of numerical variables regard each other, where outliers appear isolated, or some numerical variables regard to some categorical variables using boxplots. Boxplots with whiskers that represent one-and-a-half the interquartile range ($IQR = Q_3 - Q_1$) are very useful to display potential outliers, which are located beyond the whiskers.

Imbalanced data sets. Like other Inductive learners, classification trees are designed to minimize errors over the training examples. Thus, when one of the categories of the dependent variable is underrepresented in the training data, compared to the other classes, their few examples can be largely ignored by learning algorithms because the cost of performing well on the over-represented class outweighs the cost of doing poorly on the smaller class. There are two common solutions, but none of them are ideal. Training samples of the over represented categories can be removed, but we are losing information, which is not a good thing in a learning process. The other option is by multiplying by a factor the training samples of the underrepresented category, although this distorts artificially the relative importance of the splits for the construction of the trees. SPSS software (SPSS Inc., 2007) allows changing the weight of the samples. Overweighting the samples of the underrepresented category is exactly the same that multiplying them by a factor. The best solution could be, then, underweight those samples of the over represented categories. This also distorts the relative importance of the splits, but information is not loosed.

Incomplete data. It is an unavoidable problem in dealing with most of the real world data sources. The simplest solution is to eliminate those cases with some missing value. When many of the non-available data are from the same variable, the possibility of removing that variable can also be contemplated. Other possibility is to test both training data sets, one of them with all the variables but fewer samples, and the other one with all the cases and fewer variables, or even as well a halfway data set, compromise between them. All this is more laborious, but sometimes it is worth. It depends on the number of missing values and its distribution among the variables. In the Surmullet database, for example, the wind variable was not available for 44 out of 163 hauls. Both training data sets were tested, one of them with the 163 cases, but excluding the wind variable, and the other one including the wind variable, but using only those 119 cases for which it was available.

Another possibility is to fill non-available data according to the rest of data, which can be applied using different criteria. Using classification trees (which dependent variable is categorical), missing data of categorical independent variables can be filled with the most common class within the same category of the sample. For numerical variables with missing data, the simplest option is to use the mean value of this variable, computed from all samples belonging to the same category of the dependent variable.

A more sophisticated alternative is to develop a regression or classification model based on complete case data for a given feature, treating it as the outcome and using all other relevant features as predictors. It has sense because very often there is some type of relationship among the independent variables. In our case of study, the mean depth of some hauls could not be calculated directly from the starting and ending point depth because one of them was not available. We used then a stepwise multiple regression analysis in order to look for some other numerical variables that combined with the available data of depth allowed to estimate the mean depth. Linear combinations of the latitude and longitude, the duration of the haul and the available data of depth, allowed a little better estimation of the mean depth than the starting or ending point depth alone.

There are even two possibilities more. One of them is to use the value of the most similar case to the case with the missing value, and the last one, only useful for categorical variables, is to create a new category as "unknown".

It is not easy to establish which of these methods has to be chosen, and sometimes it is worth to test some of them.

Discretization. Unlike multivariate regression or discriminant analysis, for regression and classification trees, numerical variables finally act as discrete ones. The simplest discretization method calculates the maximum and the minimum for the variable that is being discretized and partitions the range observed into k equal sized intervals. The range can also be partitioned into intervals containing the same number of instances. Sometimes it is worth to represent the variable to look for gaps which can be used to establish the intervals. Some variables can be partitioned according to logical criteria or with some type of sense.

SPSS 17.0 software only allows the equal-sized-intervals way of discretization. It allows choosing the number of intervals for each numerical variables. In our case study, different intervals were tested, and some quantitative variables were divided according to some logical criteria. For example, the hour of the mean point of the hauls, which ranged from 5.30 to 15.30, was partitioned in early morning (before 8.00), morning (between 8.00 and 12.00) and afternoon (after 12.00). The day of the year (1-365), for example, was discretized in months and seasons.

There are many ways of discretizing different types of variables. Latitude and longitude are numerical variables, but sometimes, samples are located in more or less defined areas, and both variables can be simultaneously partitioned in area 1, 2, etc. Wind direction (0-360°) can be partitioned in different components (i.e., N, NW, W, etc.).

There are also some complex discretization methods that only will be mentioned here. One of them, proposed by Maas (1994) evaluates candidate cut points against an error function and explores a search space of boundary points to minimize the misclassifications on the training set. Fayyad and Irani (1993) selects the cut-points minimizing entropy until a stopping criterion, based on the Minimum Description Length criterion, ends the recursion. Finally, Boulle (2004) proposes a method based on the global optimization of chi-square. It is a Bottom-up method, which starts from the set of single value intervals and iteratively merges neighboring intervals (Kotsiantis et al. 2006).

Application of the Induction Algorithms

Software for implementing CART, including both free open source as well as commercial programs, has been available for many years. These software products differ with respect to the available induction algorithms and their modelling capabilities. Graphical and statistical diagnostics, procedures for validation, and methods for controlling for overfitting, vary across them. In both cases of study, analyses were performed using SPSS 17.0, although there are many software programs freely available, which modelling capabilities are even superior. Open source software programs freely available include, for instance, Weka, Sipina, RapidMiner and R software.

Splitting criteria. Tree building begins at the root node, which includes all samples in the training dataset. Beginning with this node, the CART software finds the best possible variable to split the node into two child nodes. In order to find the best variable, the software checks all possible splitting variables, as well as all possible values of the variable to be used to split the node. In choosing the best variable, the program seeks to minimize the average "impurity" of the two child nodes, an indication of the relative homogeneity (the inverse of impurity) of cases in the nodes. If all cases in each terminal node show identical values, then node impurity is minimal, homogeneity is maximal, and classification of the cases used in the computations is perfect (which does not involve an optimal predictive capacity over new cases). A number of different measures of impurity can be used, loosely called "splitting criteria". There are three main types of splitting criteria: GINI, Entropy, and Chi-square.

The *GINI index* (t) is the measure of impurity most commonly used. It is a measure of statistical dispersion that has also found application in the study of inequalities in disciplines as diverse as economics, health science, ecology, chemistry or engineering.

$$(t) = \sum p\,(j/t)\,p(i/t)$$

Entropy (H) favours balanced or similar splits. Although the concept of entropy was originally a thermodynamic construct, it has been adapted in many other fields of study, including information theory, psychology, economics and ecology (i.e. the Shannon index of biodiversity). It is, for example, the criterion in algorithms C4.5 and ID3.

$$\mathrm{H} = -\sum p_i \,\mathrm{Ln}\, p_i$$

Chi-square measure is essentially a test of the goodness of fit (Agresti, 2002). It is similar to the standard Chi-square value computed for the expected and observed classifications. For regression trees, a least-squares deviation criterion is also used. Different induction algorithms are based on different splitting criteria, and software programs differ with respect to the available induction algorithms. Both cases of study were performed using SPSS 17.0 software. Four induction algorithms for classification are available in this program: (i) CHAID (Chi-squared automatic interaction detection), (ii) exhaustive CHAID (an improvement version of CHAID), (iii) QUEST (quick, unbiased, efficient statistical tree), and (iv) CRT (classification tree and regression). The first three approaches are based on the significance of the Chi-square value (Thomas and Galambos, 2004), and GINI is the default rule in CRT. Since the three first algorithms are all based on chi-square statistics, only exhaustive CHAID and CRT were applied, and other algorithms like C4.5, C5.0 and J48 were tested using other programs like See5, Weka, RapidMiner and Sipina.

Determining when to stop splitting. The second step in CART is to determine when to stop splitting a node and declare it a "leaf". Using an algorithm based on the GINI Index, like CRT, this is carried out establishing the minimum change in improvement for a splitting. Large values tend to yield smaller trees. When using an algorithm based on chi-square, like CHAID, splitting is controlled by means of the significance level (maximum p-values) for splitting in the partition nodes. This time, large values tend to yield larger trees. SPSS 17.0, in addition to these ways, splitting can also be controlled determining the minimum number of cases necessary in both parent and child nodes to be split, as well as the maximum depth of the trees, this is, their maximum number of levels of division.

Overfitting. If the number of splits is too high (or the samples in the terminal nodes too low) in relation to the number of cases, induction algorithm very probably will be overfit. Like in statistics, overfitting happens when a model is excessively complex and explains random error or noise instead of or in addition to the real relationship between dependent and independent variables. This occurs when the number of features involved by the tree model is relatively large compared to the training set. Despite being very accurate in fitting training data, if a tree model is overfit, it will fail in predicting new data, and wrong conclusions can be inferred from this tree. A tree model with fewer splits will be always more robust. The issue is to get a trade off between the accurate in fitting training data and the number of splits.

Pruning. It is a way of reducing overfitting, analogous in spirit to the stepwise backward approach in discriminant and multiple regression analysis. The issue is to overgrow the tree until all training samples are correctly classified, and then to prune this fully-grown tree. This pruning is guided by the classification error rates estimated on a set of samples not used for the construction of the tree (Quinlan, 1987). In data-poor problems, however, the training data can be too few if a portion is separated for the pruning process (Esposito et al. 1997). An alternative to this procedure is the cost-complexity pruning, where the goal is to minimize a function that considers both misclassification costs and some measure of the tree complexity.

Cross validation. The best way to test the validity of a model is with a separate test sample. When data are insufficient, however, cross validation is a good alternative. The training sample is divided into *n* roughly-equal parts, each containing a similar distribution for the dependent variable. CART takes all the data less one of these n parts, constructs the largest possible tree, and uses the excluded part to obtain estimates of the error rate of selected sub-trees. The same process is then repeated (growing always the largest possible tree) on each n test samples (always estimating the error rate of selected sub-trees), until each part of the data has been held in reserve one time as a test sample. The results of the n mini-test samples are then combined to form error rates for trees of each possible size; these error rates are applied to the tree based on the entire learning sample.

Adaptive boosting. It is a different approach in the use of CART, proposed by Freund and Schapire (1996). The notion is to generate many trees, rather than just one. Each tree pays special attention to those cases misclassified by the other trees. The process continues for a predetermined number of times, generating the same number of trees, each one misclassifying a number of samples. When a new sample is analysed, each tree predicts a category. The number of predictions for each category is counted to determine the final classification. In this fashion, a result is obtained that takes into account many different features related with the difference between the categories. This approach is not available for SPSS, and was not used in our cases studies, but can be useful for some kind of problems for which, the classification of new samples is important, and many independent variables seem to be important.

In our cases of study, the procedure followed was always the same: (i) to generate several trees, (ii) to analyze each tree in detail and to compare them with each other, (iii) to obtain some conclusions about the system, and (iv) to select a minimum number of trees that represent such conclusions. In order to obtain different trees, different induction algorithms (with different splitting criteria) were applied. In each case, different partitions of the numerical variables were tested, with and without pruning and cross validation. Some trees were obtained, in addition, after removing certain variable, which played an important role in a previous tree, or forcing the entry of a given variable.

Interesting trees often cannot be represented and explained for reasons of space, so those of them that better represent the main conclusions have to be select.

Interpretation of the Trees and Knowledge Discovery

Data mining and knowledge discovery are probably the main use that CART can provide to the fishery research. Both things are the direct result of the analysis and interpretation of the trees.

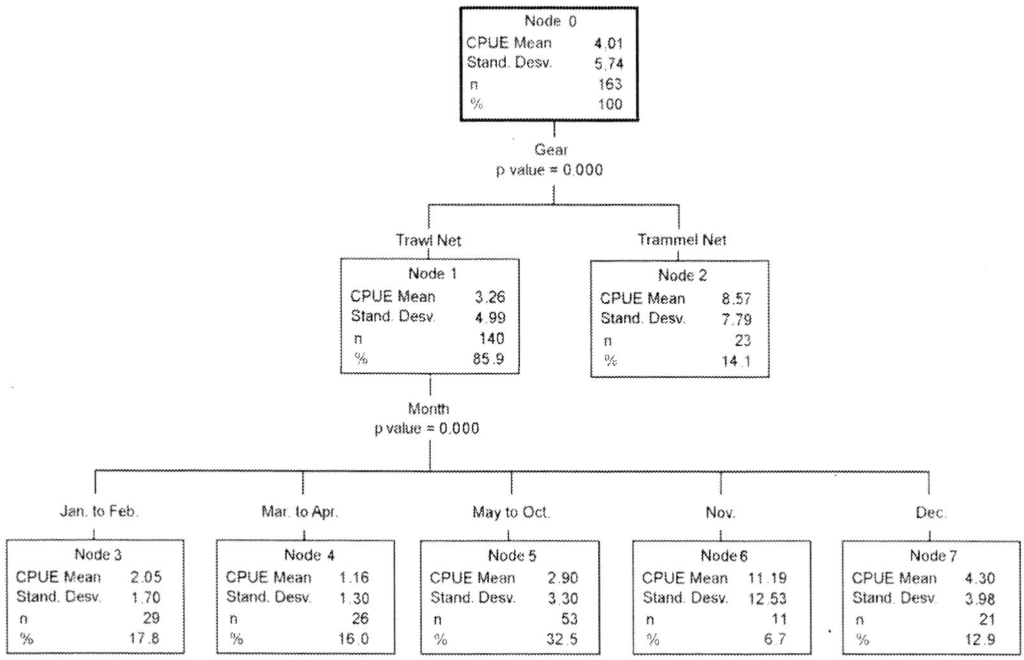

Figure 2. Tree 1 obtained with the exhaustive CHAID algorithm, the full data set and 9 equal-size intervals for the numerical variables.

Figure 2 shows tree 1, obtained in our case of study about the factors that determine the CPUE (Catch Per Unit Effort, kg/h) of the Surmullet. The first conclusion is that the gear type is a decisive factor. The CPUE mean value of the 23 hauls using trammel net (8.57, node 2) was more than twice the mean of the 140 hauls using trawl gear (3.25, node 1). None of the remaining variables had a significant effect on the CPUE for the 23 hauls using trammel net, despite their high standard deviation (7.79, node 2). This could be because these variations are not related to any of the remaining independent variables, although more probably, it is just because the number of samples (n=23, node 2) is not large enough. For the 140 hauls using trawl gear (node 1), however, the variable month was very significant ($p<0.0001$). According to tree 1, the highest CPUE mean value using trawl gear was on November (11.19, node 6), followed by December (4.30, node 7), and the minimum in March and April (1.16, node 4).

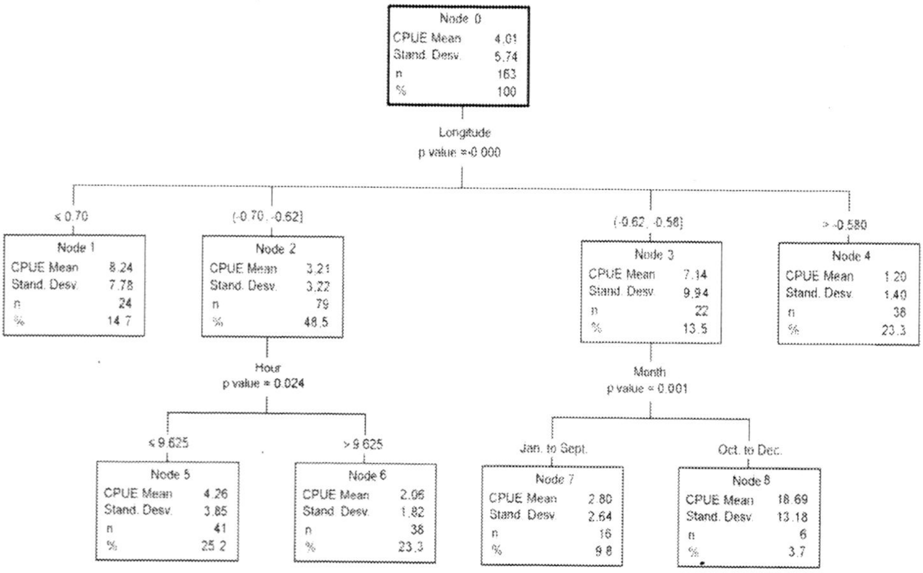

Figure 3. Tree 2 obtained with the exhaustive CHAID algorithm and the full data set and 13 equal-size intervals for the numerical variables.

In SPSS 17.0, using the exhaustive CHAID algorithm, the default number of equal-size intervals in which numerical variables are partitioned is 10. The tree obtained with this number was not very conclusive, and different partitions were tested. Tree 1 was obtained with 9 intervals, but using 13 intervals, the result was the tree 2 (figure 3). The variable longitude became more significant than the gear type, which lost its significance. This indicates that there is some type of correspondence between both variables. Tree 2 shows that this is the case, so most of the samples included in the first interval (node 1, longitude < -0.70) are from trammel net (see also figures 3 and 4). The correspondence is so definite[1] that, at first, it is not possible to resolve which variable is the decisive factor.

Trammel net is the main gear used when the target species is the Surmullet, so very probably the gear type is a key factor. However, tree 2 also shows that there are important differences among the other three intervals of longitude. The 22 samples located in the thin band from -0.62 to -0.58 have the highest CPUE mean value (7.14, node 3). This value is lower in the 79 samples placed to the west side of them (3.20, node 2), and much lower in the

[1] This clear-cut is related to the limits that the Spanish law establishes in this area to fish with trawl gear (Fernández, 2009).

remaining 38 samples (1.20, node 4) that occupy the wide area situated to the east side of the area (see figures 4 and 5).

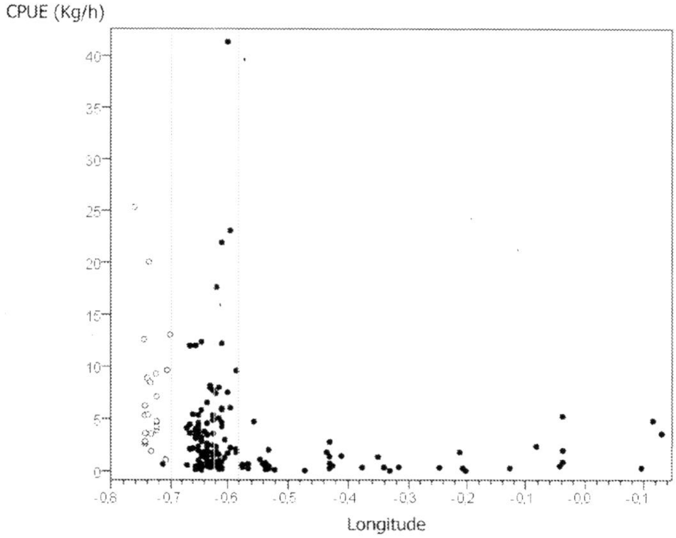

Figure 4. Bivariate plot showing the relationships between the CPUE and the longitude and position of the limits between nodes 2, 3 and 4 of tree 2. Samples from trammel net (empty circles), samples from trawl net (black circles).

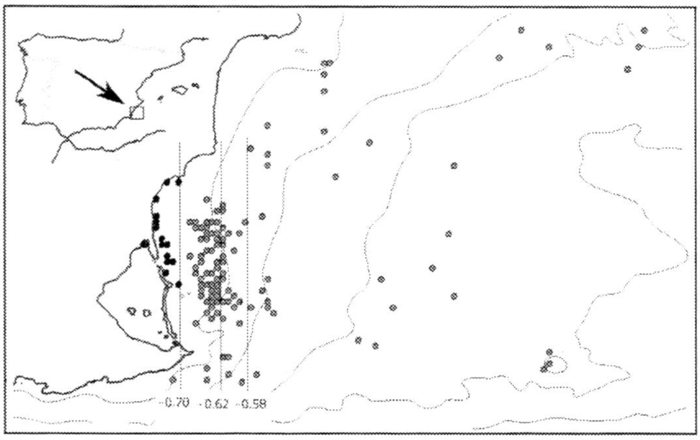

Figure 5. Area of study with location of the hauls and position of the limits between nodes 2, 3 and 4 of tree 2.

The variable longitude explains where the CPUE of the Surmullet is higher or lower, but not the ecological factors that determine it, so posterior analyses were performed excluding the coordinates. Both types of gear are not fully comparable, given that they work in a very different way, so the 23 samples from trammel net were excluded as well.

Tree 3 (figure 6) is a combination of two trees obtained as well with the exhaustive CHAID algorithm, but without the variables longitude and latitude, and the 23 samples from trammel net. It allows a quite clear identification of some key ecological factors affecting the captures of the Surmullet. The CPUE is primarily affected by the time of year, and later by the depth. Secondarily, the CPUE is affected significantly by the hour of the day in some conditions. The highest CPUE mean value is obtained in autumn (5.70, node 2), which agree with tree 1. Part of the CPUE variations among the autumn samples is explained by the depth. The 6 hauls comprised between 50 and 77 meters show the highest CPUE mean value (17.53, node 9). Below (until 18 meters) and above (until 473 meters) the mean values are much lower (respectively 3.89, node 8 and 4.17, node 10).

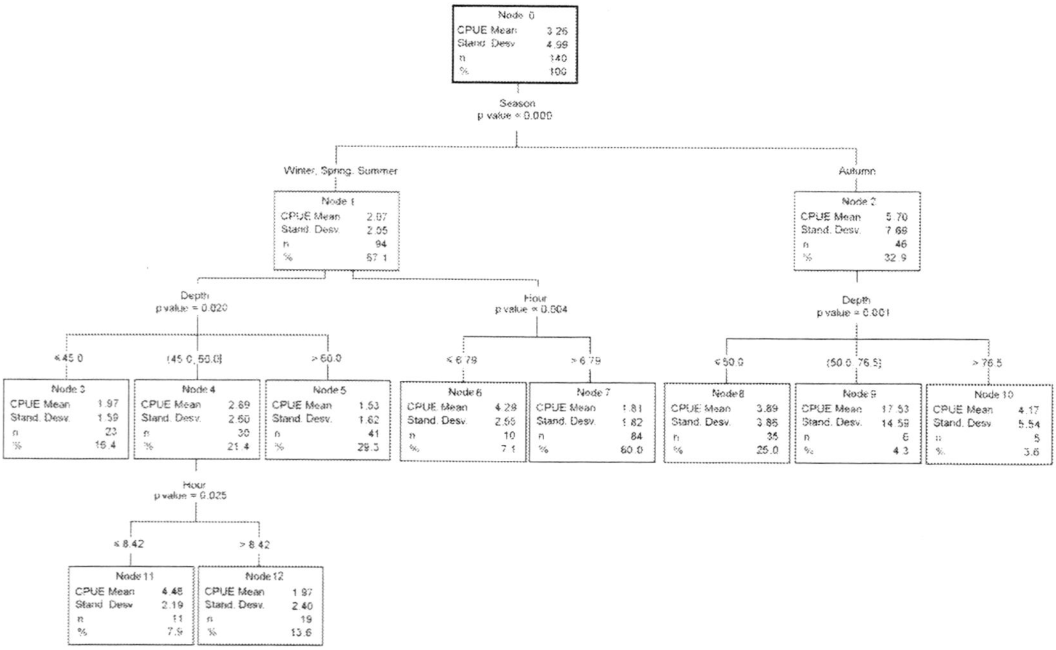

Figure 6. Tree 3 was built from the combination of two trees obtained with the exhaustive CHAID algorithm, without the variables longitude and latitude, and the 23 samples from trammel net.

Tree 3 shows that the CPUE mean value of the rest of the seasons (winter, spring and summer) is lower. Like in autumn, the depth also explains part of the variations of the CPUE among these samples, which can also be explained by the hour of the day. Both factors, depth and hour, appear in tree 3 in independent branches because they were taken out from two trees, which other branches are the same. Unlike in autumn, the hauls showing the maximum CPUE mean were comprised between 45 and 60 meters (2.89, node 4). Among them, the CPUE mean value is higher when the mean point of the haul was before 8.25 a.m. (8.42 in decimal scale, nodes 11 and 12). The hour of the day also affects in a significant way to the whole of the samples of winter, spring and summer, but the limit was at 6.47 a.m. (6.79 in decimal scale, nodes 6 and 7).

Tree 4 (figure 7) was obtained with a different training data set, which also includes the variable wind, but only the 119 samples for which this variable is available. It was obtained using the CRT algorithm, which is based on the improvement of the GINI index, instead of the significance of the Chi-square value. A tree, almost equal, but without the last branch (nodes 5 and 6), was obtained with the first data set when the CRT algorithm was applied. These trees confirm that the time of the year and the depth are the main factors determining the CPUE. According to tree 4, the wind also affects the capture of the Surmullet, al least on

November and December, above the 53.2 meters (nodes 5 and 6), but not during the rest of the year, or in deeper waters. With gentle breeze or higher, the mean capture is reduced to approximately half (nodes 5 and 6). Tree 4 and the similar one obtained with the first data set show that it is possible to obtain also interesting and very different trees, applying different algorithms, especially when they are based on different splitting criteria.

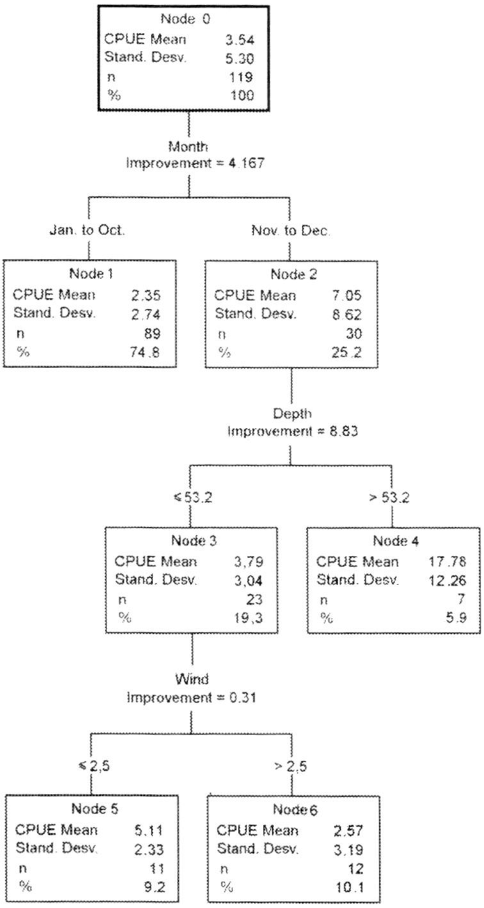

Figure 7. Tree 4 obtained with the CRT algorithm with a training data set that also includes the variable wind, but only the 119 samples for which this variable is available.

CONCLUSION

Our case of study shows

(i) that previous knowledge of the system is essential,
(ii) that direct interpretation of the tree models without a detail analysis may lead to wrong conclusions, and
(iii) that graphical representations can be an indispensable complement for a correct interpretation of the trees.

It is important to highlight that there is an enormous amount of very different problems that can be dealt with CART, each one with its own singularities. All them share the aim of extracting some knowledge on a system about which we have certain information in form of quantitative or qualitative data. The point is to prepare this information in an appropriate manner, enabling the software to learn what we like to know. If we get it, the knowledge required will be in the representation of the product of learning; this is, in the trees. Thus, the number of potential applications in fishery research and management is huge. The number of applications of CART and other techniques of machine learning, moreover, increases each year, which shows that machine learning is becoming a promising tool for fisheries management.

REFERENCES

Agresti, A., 2002. Categorical Data Analysis, Second ed.Wiley and Sons, New York.

Boulle M. (2004). Khiops: A Statistical Discretization Method of Continuous Attributes. *Machine Learning* 55:1 53-69.

Breiman, L., J. Friedman, R. Olshen, and C. Stone (1993), Classification and Regression Trees, New York: Chapman and Hall.

Esposito, F., Malerba, D., Semeraro, G., Kay, J. (1997). A comparative analysis of methods for pruning decision trees. *IEEE Transactions on Pattern Analysis and Machine Intelligence* 19(5) 476–491.

Fayyad U., and Irani K. (1993). Multi-interval discretization of continuous-valued attributes for classification learning. In *Proc. of the Thirteenth Int. Joint Conference on Artificial Intelligence*, 1022-1027.

Fernandes, P. G. (2009). Classification trees for species identification of fish-school echotraces. – ICES *Journal of Marine Science*, 66: 000–000.

Fernández, A. M. (2009). La pesquería artesanal de Santa Pola (SE de la Península Ibérica). Identificación de tácticas de pesca y estimación de índices de abundancia. Tesis Doctoral. Universidad de Alicante.

Freund, Y. and Schapire R.E. (1996). Experiments with a new boosting algorithm. In *Machine Learning: Proceedings of the Thirteenth International Conference*. Morgan Kauffman, San Francisco, pp. 148-156.

James G. A. and James, R. A. (2010). Smart data: enterprise performance optimization strategy. Wiley series in systems engineering and management.

Kotsiantis, S. B., I. D. Zaharakis and P. E. Pintelas (2006). Machine learning: a review of classification and combining techniques. *Artificial Intelligence Review* 26 (3), 159-190.

Maas W. (1994). Efficient agnostic PAC-learning with simple hypotheses. *Proceedings of the 7th ACM Conf. on Computational Learning Theory*, 67-75.

Mendoza, M. (2007). Decision Trees: a Machine Learning Methodology for characterizing Morphological Patterns resulting from Ecological Adaptations. In: *Automated Object Identification in Systematics: Theory, Approaches, and Applications*, Pp:261-276. MacLeod, N. (Ed.), Systematics Association's Special Volume Series, UK. ISBN 0-8493-8205-X.

Mendoza, M. García, T. and Baro J. (2010). Using classification trees to study the effects of fisheries management plans on the yield of Merluccius merluccius (Linnaeus, 1758) in the Alboran Sea (Western Mediterranean) *Fisheries Research* 102 (1-2): 191-198.

Olden J. D., N. LeRoy Poff, and Kevin R. Bestgen (2008). Life-history strategies predict fish invasions and extirpations in the colorado river basin. *Ecology*, 89: 847–856.

Pons, M., Marroni, S., Machado I., Ghattas B., Domingo A. (2009). Machine Learning Procedures: An Application to By-Catch Data of the Marine Turtles *Caretta Caretta* in the Southwestern Atlantic Ocean. Collect. Vol. Sci. Pap. ICCAT, 64(7): 2443-2454.

Quinlan, J.R., (1985). Introduction to decision trees. *Machine Learning*, 1: 81-106.

Quinlan, J.R., (1987). Simplifying decision trees, *International Journal of Man-Machine Studies*. 27(3): 221–234.

Rogan, J., Franklin, J., Stow, D., Miller, J., Woodcock, C., Roberts, D., (2008). Mapping land-cover modifications over large areas: A comparison of machine learning algorithms. *Remote Sensing of Environment*, 112: 2272-2283.

Thomas, E. H., and Galambos, N. (2004). What satisfies students? Mining student-opinion data with regression and decision tree analysis. *Research in Higher Education,* 45: 251-269.

Simmonds, J. E., Armstrong, F. and Copland, P. J. (1996). Species identification using wideband backscatter with neural network and discriminant analysis. *ICES Journal of Marine Science*, 53: 189–195.

SPSS Inc. (2007). SPSS 16. [Computer software and manual]. Chicago, IL: The Author.

Srinivasu, P., Avadhani, P.S., Korimilli, V., Ravipati, P. (2009). Approaches and Data Processing Techniques for Intrusion Detection Systems, IJCSNS *International Journal of Computer Science and Network Security*, 9 (12): 181-186.

Wang SS-E (2005) Groundfish habitat associations from video survey with a submersible off the Washington State coast. Thesis, University of Washington.

In: Fishery Management
Editor: Janice S. Intilli
ISBN: 978-1-61209-682-7

Chapter 5

APPLICATION OF SUSTAINABLE DEVELOPMENT INDICATOR SYSTEM (SDIS) TO TAIWAN DISTANT WATER FISHERIES MANAGEMENT

Hao-Tang Jhan, Wen-Hong Liu*, Kun-Lung Lin, and De-Lu Ding
Department of Fisheries Production and Management, National Kaohsiung Marine University, Nanzih District, Kaohsiung City, Taiwan

ABSTRACT

This study is concerned with the use of the Sustainable Development Indicator System (SDIS) to evaluate the performance of Taiwanese distant water fisheries management and related management policies. The four-dimension framework and factor-analysis method are used to aggregate indicators in the SDIS. The ecological index tends towards sustainability in term of exploitation rate and fisheries resources, but the fisheries statistical and biological sampling date are required to assess fish stock more accurately and helpful for set up management measures . The aging of fishing vessels age structure, cost increase and return decrease mean that the economic index tends towards unsustainability. Thus, it is suggested that the Fisheries Agency should formulate policies to increase profits for fishing, improve the operation conditions, and maintain the catch. Due to the rate of foreign fishing crews raised, employment opportunity and social performance have declined, thus the social index tends towards unsustainability. The problem of lack of domestic fishing crews remains. Since the implementation of numerous distant water fisheries management measures in 2006, the institutional index has gradually been moving towards sustainability. Although the management capacity and effectiveness of Taiwanese distant water fisheries has been improving markedly, it is necessary to devote further attention to the problems of management efficiency and cooperation. Lastly, despite the objective of reducing the number of Taiwanese distant water fishing vessels, but the Fisheries Agency has failed to reduce the number of old fishing vessels leading to unsustainability of the fleet due to increasing cost and decreasing profilts.

* Address: No.142, Haijhuan Rd., Nanzih District, Kaohsiung City 81143, Taiwan (R.O.C.). Tel: +886-7-3617141 ext. 3528. Fax: +886-7-3642297. Email address: andersonliu@mail.nkmu.edu.tw.

Keywords: Taiwanese distant water fishery management, four-dimension framework, factor-analysis method, index

INTRODUCTION

Taiwan's fisheries developed and increased production after World War II. However, a number of have problems caused a decline production of both offshore and coastal fisheries in recent year. For example, pollution, habitat destruction and overfishing. For the goal of encourage development of distant water fisheries and construction of large fishing vessels, Taiwan's government introduced a distant water fisheries development program in 1979 (Luo, 1998). The production of Taiwan fisheries increased from 0.25 million tonnes in 1960 to 1.34 million tonnes in 2008; landed value also increased from $NT 24.6 hundred million to 920.58 hundred million. The production and value of distant water fisheries increased 62% and 47% respectively illustrating the importance of distant water fisheries for Taiwan.

Taiwan distant water fisheries include Pacific, Atlantic and Indian Ocean components of main distant water fisheries in the world, especially for tuna and squid fishery (Hu, 2004). As the emphasis in international fisheries management progressed from utilization to conservation of resources and the role of Regional Fishery Management Organizations (RFMOs) increased, the Fishery Agency introduced several measures to monitor fishing effort and enhance management effectiveness ecommendation 05-02 by the International Commission for the Conservation of Atlantic Tunas. Currently Taiwan does not apply sustainable fisheries development indicator system (SFDIS) to assess distant water fisheries and most indicator research aimed at developing indicator technology does not research in the relationship between results and decision-making (Potts, 2006). Accordingly, this paper considers the role of a Sustainable Fishery Development Indicator System (SFDIS) that covers ecological, economic, social and institutional dimensions to assess the performance of Taiwan's distant water fishery management with appropriate reference to FAO Technical Guidelines for Responsible Fisheries No.8-Indicatros for sustainable development of marine capture fisheries (1999) and the background of Taiwan distant water fisheries.

THE STATE OF THE DISTANT WATER FISHERIES

In recent years, the production of distant water fisheries has fluctuated considerably, being over 0.7 million tonnes per year since 1997. The dramatic changes have especially been influenced by the squid fishery. In 2008, the main distant water fisheries production was squid, tuna longline and tuna purse seine fisheries respectively. The value of distant water fisheries has been maintained at more than $NT40 billion since 2000 and over 50% of the value is the tuna longline fishery.

The number of people employed in distant water fisheries has decreased from 33,162 in 1995 to 17,396 in 2008. This has been largely due to a decrease in the number of full-time fishermen from 31,970 in 1995 to 16,783 in 2008. The number of fishing craft in distant water fisheries has declined from 1,672 in 1990 to 837 in 2008. The number of tuna longline vessels has decreased the most, by 390 vessels between 1990 to 2008; followed by squid and tuna purse seine vessels.

The species of fish found in fish catch in distant water fisheries may be divided into four principal types: squid, tuna, skipjack tuna and Pacific saury. The most valuable of distant water fisheries is tuna, especially bigeye and yellowfin tuna.

PROBLEMS

The problems of the distant water fisheries may be classified on ecological. Economic, social and institutional dimensions in the SFDIS (Table 1).

Table 1. Selection of indicators for a sustainable indicator system to monitor management of Taiwan's distant water fisheries

Dimension	Issues	Indicators
Ecological	Excess fishing effort	CPUE Fishing effort Landing
	By-catch	Ratio of shark by-catch
	Decline of tuna resources	Composition rate of main fish species Tuna production
Economic	Increase of operational cost	Operating cost Average price of globe crude oil Age structure of fishing vessel Unit labour of production Unit cost of production
	Decrease of return	Value Net return Profit rate Average price of yellowfin tuna The value of output
Social	Lack of fishing labour	Total employess Frequency of fishing operation accidents The number injured fishermen Ratio of average fishermen's average wage to average civil wage
	Lack the concept of international fisheries management	The number of fishing vessels detained The number of fishing crew completing safe and law courses and obtaining certification
	Excessive rate of foreign crew	Ratio of foreign crew Ratio of national crew

Table 1. (Continued)

Dimension	Issues	Indicators
Institutional	ICCAT imposed sanctions against Taiwan	Tuna catch quota
	Lack management capacity	Budget for fisheries Management The number of reduce fishing vessels The number of suspend fishing vessels The number of valid license The number of patrol vessels in high sea The number of observers The number of observers dispatched The number of observation days The number of dispatch vessels to observed Budget for observers program Ratio of observers to total vessels
	Decrease fishing area	The number of countries engaged in fisheries cooperation The number of vessels engaged in fisheries cooperation The number of participation in international congress and scientific conference of regional fisheries management organisations The number of participation in international fisheries management organisations

The Ecological Dimension

Due to the increasing number of vessels, modernisation and improved fishing technology has led to significant increase in fishing efficiency. Overcapacity is widely recognized as a major problem affecting fisheries on national levels. Before 2005, the number of distant water fisheries included 1,167 vessels and 521,000 tons in Taiwan (Fisheries Agency, 2009). The problem of excessive fishing effort led to overfishing, fishing down marine food webs and fish stocks decline(Christensen, 2000 ; Schiermeirer, 2002 ; Pauly et al., 2005). For example: FAO(2007) revealed that only four of the 23 stocks of the principal market tunas are moderately exploited. In 2008, Taiwan tuna production from distant water fisheries has decreased from a peak in 2002 of 270,370 tonnes to 162,553 tonnes, the situation was similar to the state of global tuna resources (Fisheries Agency, 2009).

The distant water trawl fishery has the highest rate of bycatch in Taiwanese distant water fisheries (Chou et al., 2004), while the tuna longline fishery has a bycatch of seabirds, sea turtles, marine mammals and sharks, causing many disputes, especially in the issue of sharks bycatch. According to WWF statistics (2009), the global tuna longline fishery bycatch includes 250,000 sea turtles, 300,000 seabirds and 3,300,000 sharks each year. In 2002, the shark bycatch reached 20,521 tonnes and attracting censure from wildlife conservation groups and fishery resources management oragnisations regarding the issue of shark finning in Taiwan (Chuang, 2004).

The Economic Dimension

Resource decline and operating cost increases have led to the cost of distant water fisheries rising rapidly. Taiwanese distant water fisheries often use fishing access fees, which are paid by the operators to the governments of coastal states. However, the coastal states have increased the limits and access fees to conservation and management resources in their EEZs, thereby increasing the cost of fishing operations (Chang, 2004).

The average age of fishing vessels is over 15 years and the situation of ageing fishing vessels has also led to cost increases. In addition, the price of global fuel has increased amounting to between 70% to 90% in total cost in recent years, but the fish price does not increase sufficiently to compensate, raising the prospect of bankruptcy for distant water fishing industries (Ou, 2008 ; Wu, 2008).

The Social Dimension

The impacts of profit decline on crew wages has led to bonuses gradually reducing. This, coupled with poor working conditions and higher risk has led to crew shortages. Japan and South Korea are faced with the same problem but Taiwan has the advantage of employing a large number of mainland Chinese crew (Haward, 2000). However, also many fishing skippers from Japan and South Korea work in Taiwanese distant water fisheries, which will likely lead to transfer of technology and experience, and leading in term to lack of locally based skippers and crews (Sha, 2007).

Taiwanese distant water fishing vessels are often detained by coastal states. In 2004, the number of Taiwanese fishing vessels detained was 29 first declining gradually followed by a growing trend at the last three years (OFDCRC, 2010).

The Institutional Dimension

The fishing grounds of distant water fisheries lie beyond Taiwan's Exclusive Economic Zone (EEZ). It is not possible for Taiwan itself to properly manage there fisheries, due to limited availability of observers and budget and lack of patrol vessels and monitoring capacity (Wu, 2006). Instead government policy has concentrated on reducing the number of fishing vessels through implementation of a vessel buyback program. However, in the system of gross register tonnage transferred, the vessel owners were allowed to build larger vessels the using the GRT from small vessels, leading to the number of fishing vessels being reduced together with increased fishing effort (Haung, 2010).

Limited tuna quota allocated by the RFMOs and overcapacity have led to the use of flags of convenience (FOC) and illegal, unregulated, and unreported (IUU) fishing to increase income and profits. This illegal fishing and exceeding the quota for tuna longline fisheries was reported by Japan. In 2005, the International Commission for the Conservation of Atlantic Tunas (ICCAT) adopted Recommendation 05-02 to decrease the bigeye tuan quota of Taiwan from 16,500 tonnes to 4,600 tonnes and introduce further management measures (ICCAT, 2006). In addition, coastal states are introducing strict control measures to protect coastal fishery resources with in their EEZs. This, together with the difficult international diplomatic position of Taiwan makes fisheries cooperation with coastal states more difficult. The states of engage in fisheries cooperation with Taiwan decreased form a peak of 31 countries in 2001 to 24 countries in 2008, making the impact on the distant water trawl fishery more serious (OFDCOC, 2010).

ESTABLISHMENT OF A SUSTAINABLE FISHERY DEVELOPMENT INDICATOR SYSTEM

Goals and Scope

The SFDIS is divided into ecological, economic, social, and institutional dimensions. The goal of ecological sustainability is to reduce overcapacity and meet the demands of international fisheries management to achieve sustainable development. Economic sustainability focuses on maintaining the benefits and competitiveness of Taiwanese distant water fisheries. Social sustainability is aimed at increasing recruitment of fishermen, improving safety, and compliant with the international law of the sea. Institutional sustainability necessitates improving management efficiency and monitoring capabilities, including reduction of illegal fishing.

Framework of the Indicator System

Fisheries management measures have been mainly concerned with single species so far, especially for target species. But successful fisheries management should include a broader range of objectives (Garcia and Cochrane, 2005 ; Cunningham and Bostock, 2005). Those commonly used in SFDIS research at present include a four-dimensional framework (Garcia, 2000; Charles, 2001; Charles et al., 2002; FAO, 1999), ecologically sustainable development framework (Chesson, 2002; Fletcher et al., 2005; Fletcher et al., 2002) and pressure-state-response framework(Garcia, 2000; FAO, 1999; Ward, 2000). The SFDIS described in this paper utilizes the four-dimensional framework covering ecological, economic, social and institutional dimensions.

Candidate Indicator Group

To build the candidate indicator group it is necessary consider the correlation between the indicators and goals rather than just collection indicators blindly. The candidate indicators group in the present study are refer of FAO (1999), Garcia (2000), Chesson et al. (1999), Charles (2001), Liu et al. (2005) and Adrianto et al. (2005). This study adds a number of indicators in order to establish the suitable candidate indicator group of SFDIS for the distant water fisheries.

Selection of Indicators and Reference Points

Data availability is the key factor when selecting the indicators to assess sustainability of fisheries (Charles et al., 2002 ; Dahl, 2000). Thus, to selected indicators were accorded with relatedness, accessibility and acceptability (Liu and Ou, 2007). In addition, the problems of distant water fisheries in Taiwan are taken account in selecting the indicators (Table 1). The SFDIS in this study defines 20 criteria and 40 indicators within ecological, economic, social and institutional dimensions. There are 3 criteria and 6 indicators in the ecological index; the economic index comprised of 7 criteria and 10 indicators; 5 criteria and 8 indicator in the social index and the institutional index is composed of 5 criteria and 16 indicators (Table 2).

Table 2. SFDIS for Taiwan's distant water fisheries

Index	Criteria	Indicator	Unit	Sustainability
Ecological Index (Q_1)	Fisheries resources (C_1)	CPUE (I_1)	Tonne/ tonnage	↑
		Landing (I_2)	Tonne	↑
	Exploitation rate (C_2)	Fishing effort (I_3)	Tonnage	↓
	Catch structure (C_3)	Composition rate of main fish species (I_4)	%	↑
		Ratio of shark by-catch (I_5)	%	↓
		Tuna production (I_6)	Tonne	↑
Economic Index (Q2)	Cost (C_4)	Operating cost (I_7)	NT$	↓
		Average price of global crude oil (I_8)	USD	↓
	Production value(C_5)	Value (I_9)	NT$	↑
	Aging patterns of fishing vessels (C_6)	Age structure of fishing vessel (I_{10})	NT$	↓
	Economic production (C_7)	Unit cost of production (I_{11})	Tonne/ cost	↑
		Unit labour of production (I_{12})	Tonne/ person	↑
	Net returns (C_8)	Net return (I_{13})	NT$	↑
		Profit rate (I_{14})	%	↑
	Price (C_9)	Average price of yellowfin tuna (I_{15})	NT$/Kg	↑
	Output (C_{10})	The value of output (I_{16})	NT$	↑

Table 2. (Continued)

Index	Criteria	Indicator	Unit	Sustainability
Social Index (Q_3)	Employment (C_{11})	Total employment (I_{17})	Person	↑
	Workplace safety (C_{12})	Frequency of fishing operation accidents (I_{18})	Boat	↓
		The number of fishing vessels detained (I_{19})	Boat	↓
		The number of fishing crew completing safe and law courses and obtaining certification (I_{20})	Person	↑
		The number injured fishermen (I_{21})	Person	↓
	Employment opportunity (C_{13})	Ratio of national crew (I_{22})	%	↑
	Crew of non-nationals (C14)	Ratio of foreign crew (I_{23})	%	↓
	Social performance (C_{15})	Ratio of average fishermen's average wage to average civil wage (I_{24})	%	↑
Institutional Index (Q_4)	Management efficiency (C_{16})	The number of reduce fishing vessels (I_{25})	Boat	↑
		The number of valid license (I_{26})	Quantifier	↑
		The number of suspend fishing vessels (I_{27})	Boat	↑
	Management ability(C_{17})	Budget for fisheries management (I_{28})	NT$	↑
		The number of patrol vessels in high sea (I_{29})	Boat	↑
	Fisheries cooperative (C_{18})	The number of countries engaged in fisheries cooperation (I_{30})	Quantifier	↑
		The number of vessels engaged in fisheries cooperation (I_{31})	Boat	↑
		The number of participation in international congress and scientific conference of regional fisheries management organisations (I_{32})	Quantifier	↑
		The number of participation in international fisheries management organisations (I_{33})	Quantifier	↑
	Observer system (C_{19})	The number of observers (I_{34})	Person	↑
		The number of observers dispatched (I_{35})	Person	↑
		The number of observation days (I_{36})	Day	↑
		The number of dispatch vessels to observed (I_{37})	Boat	↑
		Budget for observers program (I_{38})	NT$	↑
		Ratio of observers to total vessels (I_{39})	%	↑
	Enforcement (C_{20})	Tuna catch quota (I_{40})	Tonne	↑

Because of the lack of target reference points, several studies have adopted average value or baseline year to assess the current sustainability trend of indicators (Adrianto et al., 2005 ; Chesson et al., 1999 ; Liu et al., 2005). This paper used the average indicator values from 2003 to 2008 to assess trends of indicators.

Aggregation and Exhibition

The indicator aggregation is use the factor analysis method (Liu et al., 2005 ; Lohani et al., 1984). The factor analysis is used to aggregate indicators within each category into a few common factors to simplify data and conduct trend analysis(Chang, 2001). The main types of visualization include the sustainability barometer, multi-layer kite diagram, four-sided pyramid, dashboard of sustainability, elliptical indicator cluster and compass of sustainability (FAO, 1999 ; Garcia and Staples, 2000 ; Lin and Sun, 2004 ; Yen, 2002). The kite diagram combined with the dashboard will display the changes occurred over time (Liu and Ou, 2007). Therefore, this study uses the kite diagram proposed by FAO (1999) to reveal the current potential problems.

RESULTS

Ecological Index

The fisheries resource (C_1) shows an increase from 2004 and achieves the highest value in 2007 but decreased in 2008 due to CPUE (I_1) and landings (I_2). The trend of the fisheries resource is not clearly defined. The exploitation rate (C_2) has increased since 2003 and reaching a peak in 2008 according to fishing effort (I_3). The composition rate of the main fish species (I_4), tuna production (I_5) decreased and the ratio of shark bycatch (I_6) decreased from the highest point in 2003 to the lowest point in 2008. The ecological index (Q_1) reached its lowest point of 13.91 in 2004 and increased gradually to its highest point of 88.4 in 2008 (Figure 1). Thus, the ecological dimension of Taiwan distant water fisheries is trending towards sustainability.

Economic Index

Cost (C_4), aging patterns of fishing vessels (C_6) and price (C_9)decreased from their highest point to lowest point in 2008. The trend of price was towards unsustainability due to the average price of yellowfin tuna (I_{15}) having risen slightly in 2004 and 2007, but still trended towards unsustainable development. Cost trended towards sustainability according operational cost (I_7) and average price of global crude oil (I_8). Production value (C_5) has fluctuated from 2003 to 2008. Production value decreased in 2005, 2006 and 2008 due to a decrease in value (I_9).

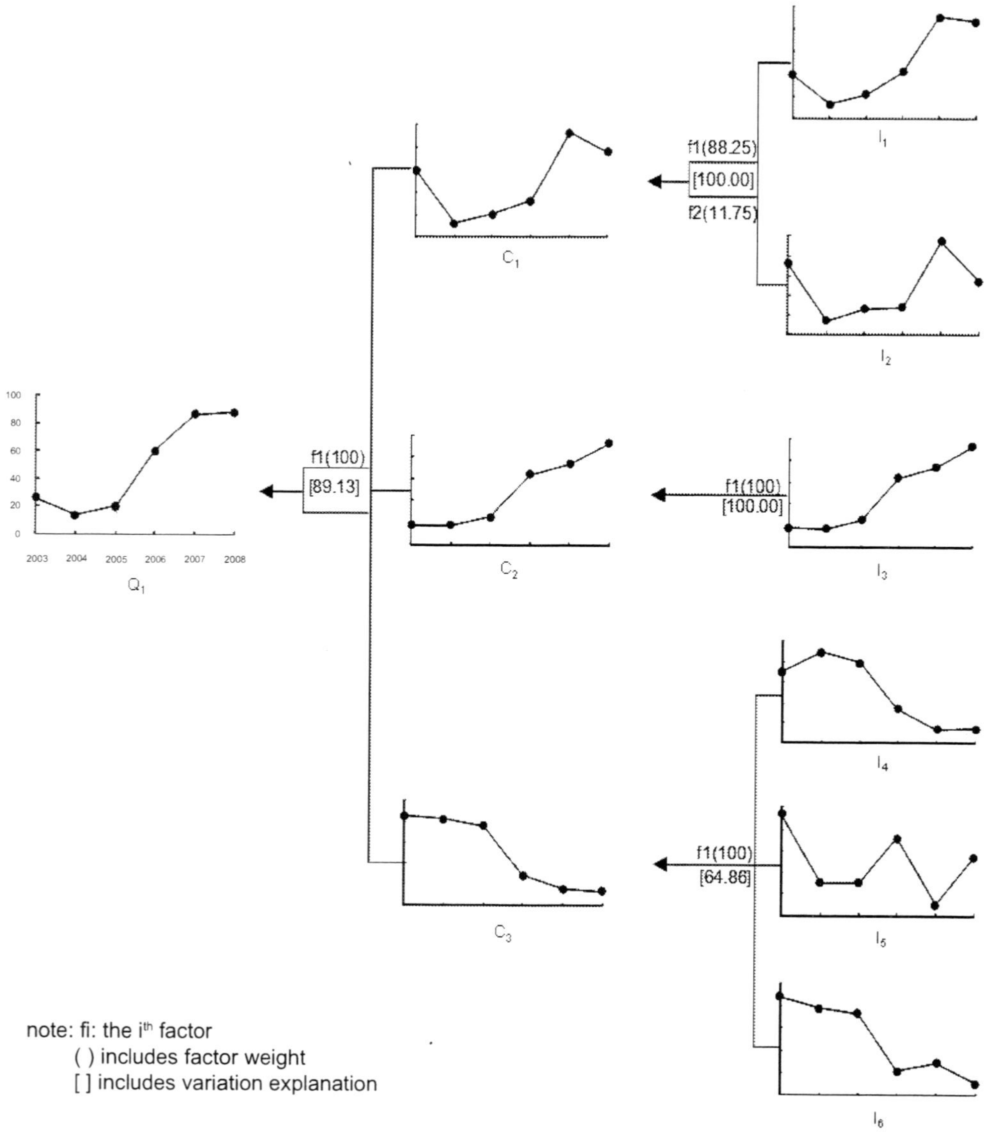

Figure 1. Process of aggregating Q_1 according to C_1~C_3.

Economic production (C_7), net returns (C_8) and output (C_{10}) were unstable with no obvious trend. In general, the economic index (Q_2) dropped from a peak of 87.96 in 2004 and increased in 2007, but it decreased and reach to the lowest point of 14.39 in 2008 (Figure 2). Accordingly, the trend of the economic index was towards unsustainability.

Social Index

Employment (C_{11}) shows a decline from 2003 and reached the lowest point in 2005, but it rose slightly form 2006 to 2008 in total employment. Workplace safety (C_{12}) decreased from 2003 to 2004, but increased from 2004 to 2008. Employment opportunity (C_{13}) and crews of non-nationals (C_{14}) decreased from their highest points in 2003 to 2008. These trend towards unsustainability due to the ratio of national crew (I_{22}) and ratio of foreign crew (I_{23}). Social performance (C_{15}) be decreased from 2004 to 2006 and increased slightly in 2007 and decreased in 2008. Generally, the social index (Q_3) has shown a steady decrease, from a maximum value of 88.92 in 2004 to a minimum value of 14.39 in 2008 (Figure 3). The social dimension in distant water fisheries in Taiwan is tending towards unsustainable development.

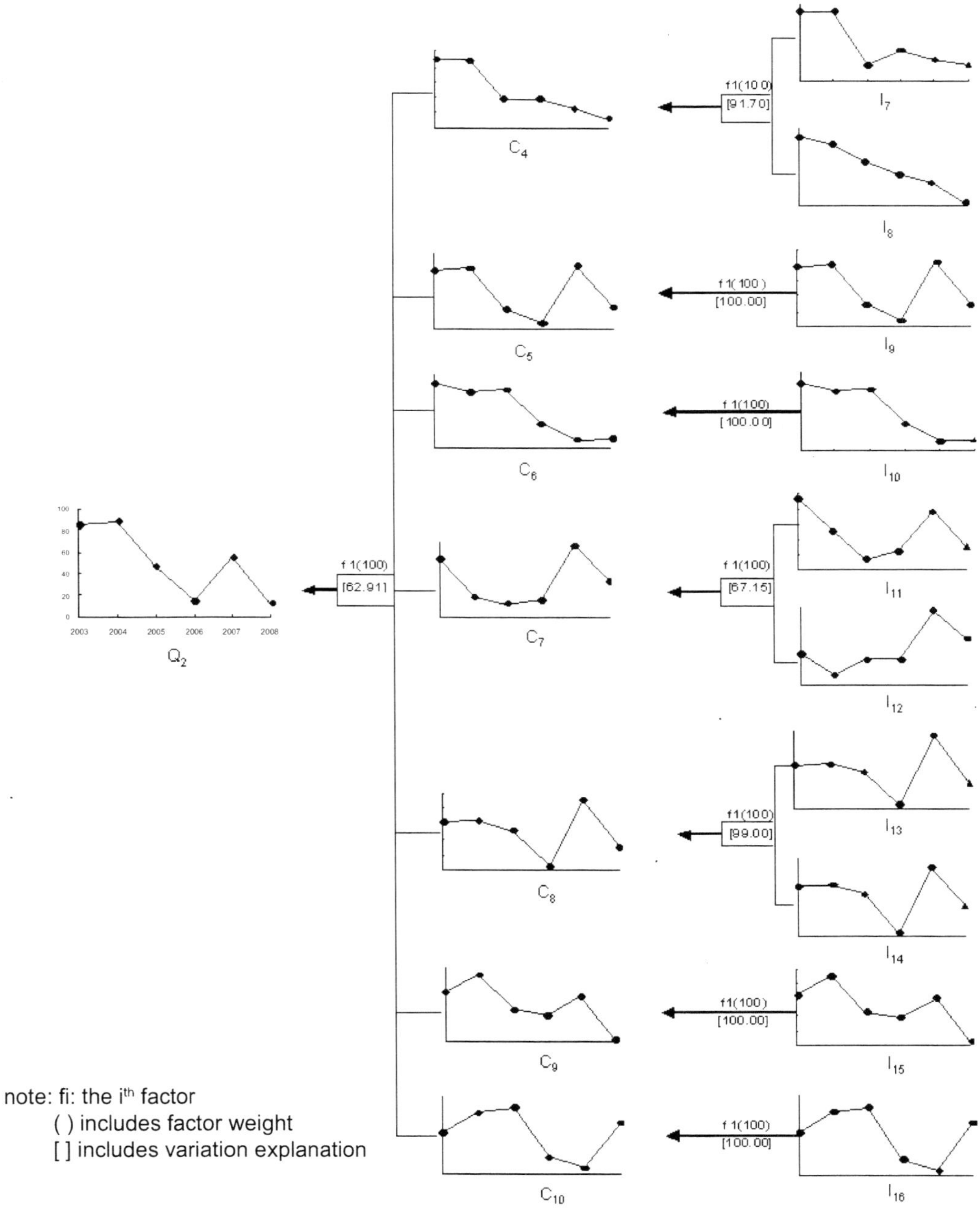

Figure 2. Process of aggregating Q_2 according to $C_4 \sim C_{10}$.

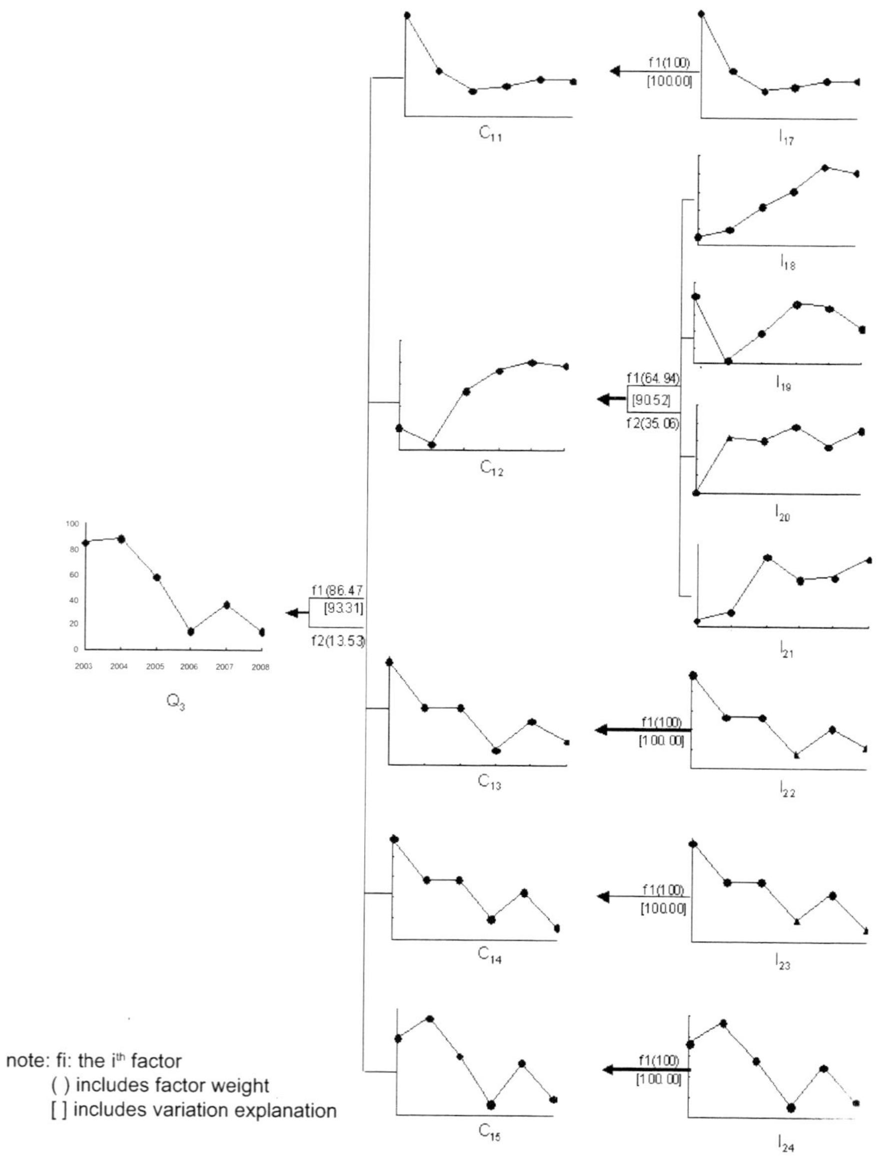

Figure 3. Process of aggregating Q_3 according to C_{11}~C_{15}.

Institutional Index

Management efficiency (C_{16}) increased from 2003 to its highest point in 2006 due to increase in the number of reduced fishing vessels (I_{25}) and suspended fishing vessels (I_{27}), but management efficiency decreased to the lowest point in 2007. Management ability (C_{17}) has increased with the increase in the budget for fisheries management (I_{28}) and patrol vessels (I_{29}) in high sea. Management ability reached it highest point in 2008 from 2005. Fisheries cooperation (C_{18}) reached its highest value in 2007 and then dropped slightly from 2007 to 2008. The observer system (C_{19}) tends to be sustainable with its lowest points in 2007 due to an increase in the number of observers (I_{34}), dispatch vessels (I_{37}) and budget (I_{38}). Enforcement (C_{20}) shows increasing trends towards sustainability from 2005 because of ICCAT reinstating Taiwan's tuna catch quota (I_{40}). Overall, the institutional index (Q_4) has been increasing and reached a peak of 89.73 in 2008 (Figure 4). Therefore, the institutional dimension in distant water fisheries in Taiwan is tending towards sustainability.

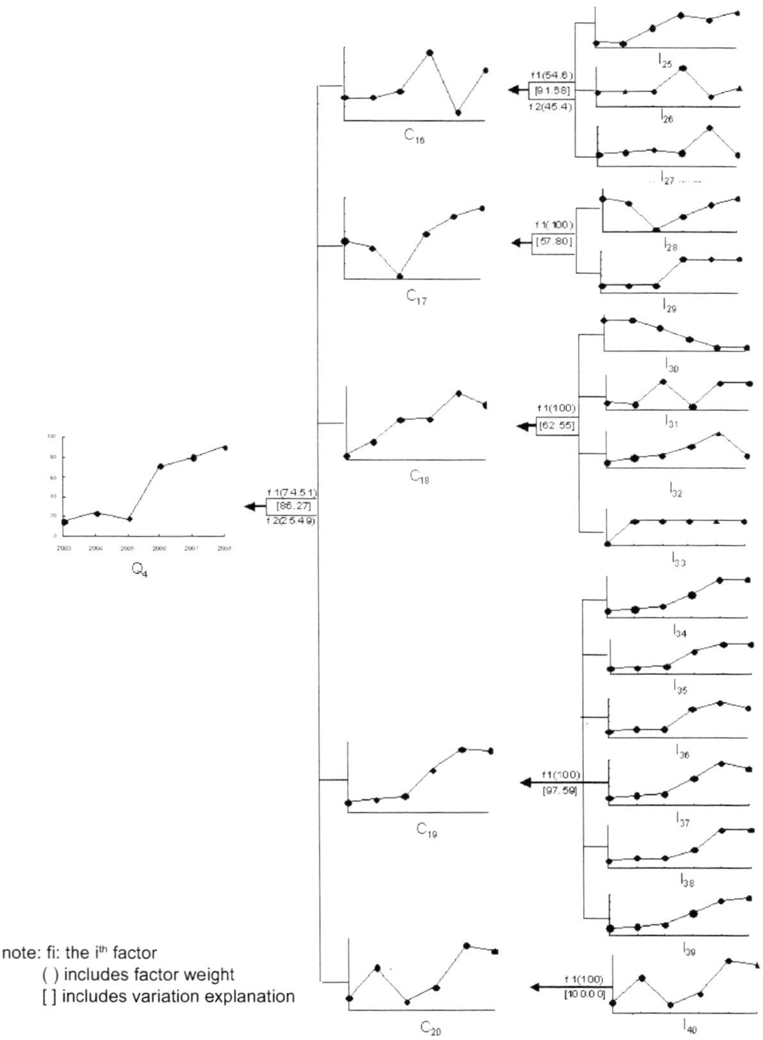

Figure 4. Process of aggregating Q_4 according to C_{16}~C_{20}.

IMPLICATIONS OF FINDINGS FOR FISHERIES MANAGEMENT

Ecological Dimension

The government implemented the policy to reduce fishing vessels (I_{25}) and suspend fishing vessels (I_{27}) after 2005. Fishing effort (I_3) decreased due to the tonnage of fishing vessels decreasing from 537,000 tonnes in 2005 to 381,916 tonnes in 2008. After reduction of the tonnage of fishing vessels CPUE (I_1) increased continuously, showing reduction of fishing effort is helpful to protect the fish resources in the high seas.

In recent years, the overall composition rate of the main fish species (I_4) and tuna production (I_6) decreased gradually, FAO (2007) and WWF (2007) are in agreement that the tuna resources have declined. Although the landing (I_2) of distant water fisheries increased from 2004 to 2006, but tuna production decreased, leading to decreased value (I_9) because of tuna has the highest economic value of target species in Taiwan. Taiwan implemented IPOS-Sharks since 2006 and the ratio of sharks bycatch (I_5) reduced significantly, but it increased again in 2007. The program requires continued monitoring and assessment.

The government implemented the policies of large-scale fishing effort in order to maintain sustainable development of fish resources. However, the various fishing grounds, fishing technology and catches of distant water fisheries make it difficult to collect physical and biological data on fish populations. Therefore, the government should request observers to assist collecting and establishing a database and to make public and exchange information with the RFMOs to improve the accuracy of assessment.

Economic Dimension

Decreases in price (I_{15}), value instability (I_9), the price of global crude oil (I_8), operating costs (I_7), and ageing of fishing vessel (I10) led to increased unit cost of production (I_{11}), net return (I_{13}) and profit rate (I_{14}) all appeared unstable trend. The distant water fishing vessels fishing effect has decreased due to the impacts of net return and profit, and the management regulations of RFMO. Taiwan tuna catch quota has been deducted by ICCAT. Although the Council of Agriculture has promoted "The Guideline of Enhance Distant Water Fishing Vessels Operations and Competitiveness" to encourage fishing, the Guideline subsidises operating cost, not contributing to increased profits, so the program does not solve the operational dilemma of the distant water fisheries.

In summary, distant water fisheries management policy in Taiwan aimed to increase production will be detrimental to fisheries sustainable development as fish resources are scarce (Haward, 2001). Therefore, Taiwan's government should establish strategies of profitable and sustainable operation for distant water fisheries.

Social Dimension

The total number of employees (I_{17}) in distant water fisheries has remained at around 18,000 since 2005. Although the ratio of national crew (I_{22}) has decreased, owners have employed more foreign crew to supplement to the work force, leading to the ratio of foreign crew (I_{23}) increasing. From the employment perspective, the ratio of national crew has reduced, but the number of employees in distant water fisheries has increased slightly which would possibly indicate actual crew numbers are lower than the statistics of the Fishery Agency. On the other hand, technological advances not only helped to reduce the frequency of fishing accidents (I18) and injured fishermen (I_{21}), but also increased in unit of labour production (I_{12}).

Although workplace safety (C_{12}) has improved, the average fishermen's wage (I_{24}) was affected the percentage of national crew. For this reason, The Fishery Agency has implemented a guidance and assistance plan to provide prizes to encourage marine-related graduate students employment on distant water fishing vessels. However, it has not achieved the intended objectives. In the future, the distant water fisheries industry may cooperate with schools to set up internship programs for training crew. In addition, for some students upcoming military service could follow alternative military service and work on fishing vessels.

Institutional Dimension

The institutional index tends towards sustainability because of government policies of reducing and suspending fishing vessels (I_{25} ; I_{27}), increasing fisheries management budget (I_{28}), the number of patrol vessels (I_{29}), and increasing and strengthening the observer system (C_{19}). In addition, the tuna catch quota (I_{40}) deducted was recovered by ICCAT. However, the fishing industry is facing a problem with fisheries cooperation states (I_{30}) reduce gradually leading to instability in the number of annual fisheries cooperation vessels (I_{31}). In the five tuna resource conservation Commissions, Western and Central Pacific Fishery Commission (WCPFC) is only active member that Taiwan joined, resulted in quota and fishing rights limited. Taiwan has joined the preparations for the South Pacific Regional Fishing Management Organisation (SPRFMO) that will help to maintain the fishing right in these waters (Ho, 2010). In conclusion, the government should strengthen the training of fisheries negotiators and to participate in international fishery management organisations (I_{32} ; I_{33}) in order to protect fishing rights and became the main country of international fisheries management gradually (Sha *et al.*, 2009).

POTENTIAL PROBLEMS

The potential problems identified by the kite diagram are based on data for 2003-2007. While the number of tune longline fishing vessels decreased after implementing the reduction fishing vessels policy (I_{25}), but the indicator of age structure of fishing vessels (I_{10}) tended towards unsustainability. That shows the government reached the target set by ICCAT, but did not to reduce number of the older vessels. Lastly, although the landing (I_2) of distant water fisheries was increased gradually, but unit cost of production (I_{11}), net return (I_{13}), and profit rate (I_{14}) has decreased since 2004 and both increased when landings reached the peak in 2007 (Figure 5).

CONCLUSION

The SFDIS is an effective measure for analyzing the problems and assessing the effectiveness of fisheries' management in sustaining fisheries' development of Taiwan's distant water fisheries. In this study, the ecological dimension tending towards sustainable development caused Taiwan's government to expend considerable effort to reduce fishing effort. But the lack of relevant biological statistical data and impacts effect in sustainability calculation.

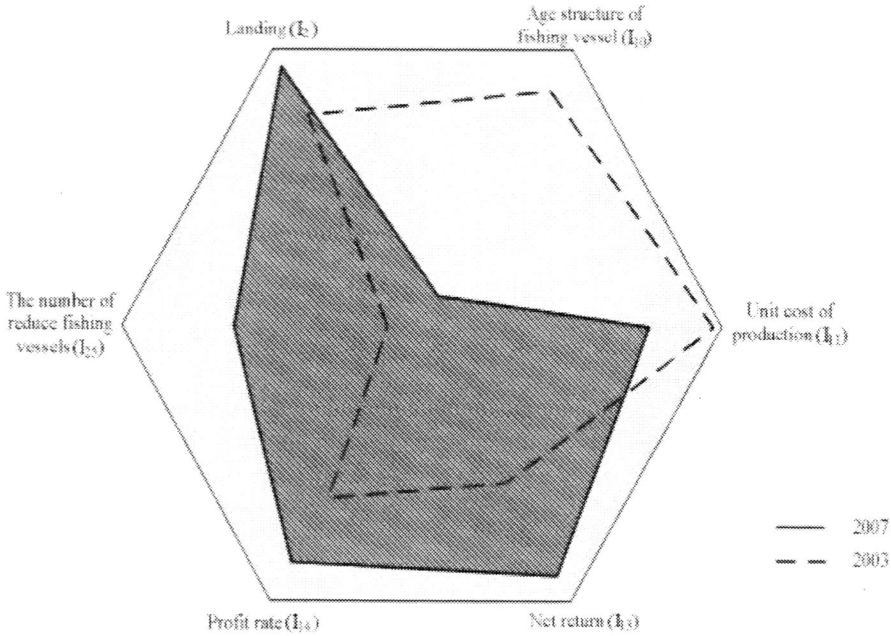

Figure 5. Dynamic kite diagram, a composite of ${}_{12}$、 I_{10}、 I_{11}、 I_{13}、 I_{14} and I_{25} (for each polygon, the value zero (0) is at the centre of the polygon and 100 is the value at the periphery of the polygon: shaded area indicates values for 2007, unshaded area indicates values for 2003).

From the result of ecological dimension, reduction in fishing effort is a positive action on recovered of fisheries' recourse. Landed catch, value, average fishermen's average wage and production per fisherman are decreasing, leading to the distant water fisheries industry being unwilling to fishing and young people are reluctant to seek work in distant water fisheries. To ameliorate this problem, government could formulate a program of profitable fisheries development to ensure the benefits of distant water fisheries and encourage industry and schools to cooperate in training fisheries personnel. In order to sustain distant water fisheries development, Taiwan's government should conform to the current trends of global fisheries management, including reduced fishing effort, instead of promoting economic interests. Lastly, Taiwan's government should change the emphasis in planning from the development production to the role of fisheries management in the future.

REFERENCES

Adrianto, L., Matsuda, Y. and Sakuma, Y. 2005. Assessing local sustainability of fisheries system: a multi-criteria participatory approach with the case of Yoron Island, Kagoshima prefecture, Japan. *Marine Policy* 29(1): 9-23.

Chang, C. S. 2004. Overview of fisheries cooperative in Taiwan tuna fishery. *Marine Kaohsiung* 3: 41-46 (in Chinese).

Chang, I. C. 2001. *Application of the Factor Analysis method for establishing a sustainable development indicator system for the island of Taiwan.* Ph. D. Thesis, University of Taiwan.

Charles, A. T. 2001. *Sustainable fishery system.* Blackwell Science Ltd, London.

Charles, A. T., Boyd, H., Lavers, A. and Benjamin, C. 2002. *The Nova Scotia GPI fisheries and marine environment accounts – a preliminary set of ecological, socio-economic and institutional indicators for Nova Scotia's fisheries and marine environment*. Saint Mary's University, Halifax N. S., Canada.

Chesson, J. 2002. *Sustainability indicators - measuring our progress*. Science for Decision Makers. Bureau of Rural Science. Barton, Australian.

Chesson, J., Clayton, H. and Whitworth, B. 1999. Evaluation of fisheries-management systems with respect to sustainable development. *ICES Journal of Marine Science* 56: 980-984.

Chou, Y. H., Kuo, C. T. and Huang, S. T. 2004. Fisheries bycatch and Regulation in Taiwan. *Fisheries Research Institute Special Publication* 5: 9-10 (in Chinese).

Christensen, V. 2000. Indicators for marine ecosystems affected by fisheries. *Marine and Freshwater Research* 51: 447-450.

Chuang, S. J. 2004. Sharks bycatch of far sea fisheries tuna longline. *Fisheries Research Institute Special Publication* 5: 47 (in Chinese).

Cunningham, S. and Bostock, T. 2005. *Successful fisheries management*. Eburon Academic, Delft.

Dahl, A. L. 2000. Using indicators to measure sustainability: recent methodological and conceptual developments. *Marine and Freshwater Research* 51: 427-433.

Fisheries Agency 2010. *The 2008 Statistical Yearbook of Fisheries in Taiwan*. Kaohsiung: Fisheries Agency (in Chinese).

Fletcher, W. J., Chesson, J., Fisher, M., Sainabury, K. J., Hundloe, T., Smith, A. D. M. and Whitworth, B. 2002. *National ESD reporting framework for Australian fisheries: the 'how to' guide for wild capture fisheries*. FRDC project 2000/145, Canberra, Australia.

Fletcher, W. J., Chesson, J., Sainsbury, K. J., Hundloe, T. J. and Fisher, M. 2005. A flexible and practical framework for reporting on ecologically sustainable development for wild capture fisheries. *Fisheries Research* 71: 175-183.

Food and Agriculture Organisation of United Nations. 1999. *FAO Technical guidelines for responsible fisheries No.8-Indicators for sustainable development of marine capture fisheries*. Available at http://www.fao.org/DOCREP/004/X3307E/X3307E00.HTM (Accessed 20 Jul, 2009).

Food and Agriculture Organisation of United Nations. 2007. *Global fishery resources of tuna and tuna-like species*. FAO Fisheries Technology Paper No. 483. Roma: FAO.

Garcia, S. M. 2000. The FAO of sustainable development and the Code of Conduct for Responsible Fisheries: an analysis of the related principles, criteria and indicators. *Marine and Freshwater Research* 51: 535-541.

Garcia, S. M. and Cochrane, K. L. 2005. Ecosystem approach to fisheries: a review of implementation guidelines. *ICES Journal of Marine Science* 51: 311-318.

Garcia, S. M. and Staples, D. J. 2000. Sustainability reference systems and indicators for responsible marine capture fisheries: a review of concepts and elements for a set of guidelines. *Marine and Freshwater Research* 51: 383-426.

Haward, M. and Bergin, A. 2000. Taiwan's distant water tuna fisheries. *Marine Policy* 24(1):33-43.

Haward, M. and Bergin, A. 2001. The political economy of Japanese distant water tuna fisheries. *Marine Policy* 25(2): 91-101.

HO, S. C. 2010. Taiwan will become the member of South Pacific Regional Fishing Management Organisation. *International of the World Fisheries Information* 206: 19-22 (in Chinese).

Hu, H. H. 2004. To strengthen Taiwan's fishing industry and internationalization. *Fisheries Extension.* 210: 14-33 (in Chinese).

Huang, H. W. and Chuang, C. T. 2010. Fishing capacity management in Taiwan: experiences and prospects. *Marine Policy* 34(1): 70-76.

International Commission for the Conservation of Atlantic Tunas. 2006. *Report for biennial period, 2004–2005 part II (2005),vol.1.* Madrid: ICCAT.

Lin, H. W. and Sun, C. H. 2004. *Indicator overall evaluation model of the International Institute of Sustainable Development.* N. page. Online. FAO. Internet. Available at http://www.law.ntu.edu.tw/sustain/project/ information/104.htm (Accessed 28 Sam, 2009).

Liu, W. H. and Ou, C. H. 2007. A comparative analysis of sustainable fishery development indicator systems in Australia and Canada. *Sustainable Development* 15: 28-40.

Liu, W. H., Ou, C. H. and Ting, K. H. 2005. Sustainable coastal fishery development indicator system: a case of Gungliau, Taiwan. *Marine Policy* 29(3): 199-210.

Lohani, N., Bindu, M. ASCE and Todino, G.. 1984. Water quality index for Chao Praya river. *Journal of Environmental Engineering* 110(6): 1163-1176.

Luo, C. C. 1998. *The history of Taiwan fisheries.* Taipei: Legislative Yuan Office of Chin-Chuan, Liu (in Chinese).

Ou, C. S. 2008. Exhausted fishery recourses and costly oil price greatly impact on global fishery. *Marine Kaohsiung* 18: 23 (in Chinese).

Overseas Fisheries Development Council of the Republic of China. 2010. *Fishing Vessels Detained by Coastal States.* Available at http://www.ofdc.org.tw/englishversion/ eINDEX.htm. (Accessed 2 Jun, 2009) (in Chinese).

Overseas Fisheries Development Council of the Republic of China. 2009. *Current status of fisheries cooperation.* Available at http://www.ofdc.org.tw/englishversion/eINDEX.htm (Accessed 2 Jun, 2010) (in Chinese).

Pauly, D., Christensen, V., Guenette, S. Pithcer, T. J., Sumaila, R., Walters, C. J. Watson, R. and Zeller, D. 2002. Towards sustainability in world fisheries. *Nature* 418: 689-695.

Pauly, D., Watson, R. and Alder, J. 2005. Global trends in world fisheries: impacts on marine ecosystems and food security. *Philosophical Transactions of The Royalb Society* 360: 5-12.

Potts, T. (2006). A framework for the analysis of sustainability indicator systems in fisheries. *Ocean and Coastal Management* 49: 259-280.

Schiermeirer, Q. 2002. How many more fish in the sea? *Nature* 419: 662-665.

Sha, C. Y. 2007. *Overview and prospect of Taiwan distant water fisheries.* Taipei: Chung-Chen Agriculture, Technology, Social Foundation (in Chinese).

Sha, C. Y., Huang, S. W. and Lin, T. J. 2007. Sustainable development and international relations - the experience of Taiwan. Taipei: Taipei Research Fund.

Tsai, J. Y. 2006. Distant water fisheries policies and management. *Fisheries Extension.* 243: 12-19 (in Chinese).

Ward, T. J. 2000. Indicators for assessing the sustainability of Australia's marine ecosystems. *Marine and Freshwater Research* 51: 435-446.

Wu, H. C. 2006. The management of distant water tuna fishery in the future. *Taiwan Fishery* 638: 22-26.

World Wide Fund for Nature. 2007. *Tuna in trouble-The challenges facing the world's tune fisheries*. Gland: WWF.

Wu, G. C. 2008. The fuel expense to rise impact to tune fishery in Japan. *International of the World Fisheries Information* 182: 18-19 (in Chinese).

World Wide Fund for Nature. 2009. *Bycatch victims.* Available at http://www.panda.org/what_we_do/footprint/smart_fishing/bycatch/issue/bycatch_victims/ (Accessed 3 Sep, 2009).

Yen C. P. 2002. *Sustainable development evaluation theory and practice*. Beijing: China Environmental Science press (in Chinese).

In: Fishery Management
Editor: Janice S. Intilli

ISBN: 978-1-61209-682-7

Chapter 6

TOWARDS SMART MANAGEMENT OF THE BALTIC SEA FISHERY

***R. Aps*[*1] *and H. Lassen*[2]**
[1]Estonian Marine Institute, University of Tartu, Tallinn, Estonia
[2]International Council for the Exploration of the Sea, Copenhagen V, Denmark

ABSTRACT

Europe 2020: strategy for smart, sustainable and inclusive growth is proposing among other flagship initiatives the resource efficient Europe. Integrated Maritime Policy for the European Union (IMP) considers the Maritime Spatial Planning (MSP) as a fundamental tool for the sustainable development of marine areas and coastal regions, and for the restoration of Europe's seas to environmental health. Governance of marine spaces aims to provide a mechanism for a strategic and integrated plan-based approach for marine management that makes it possible to manage current and potentially conflicting uses. Fishery systems are an obvious example of coupled human and natural systems. A precondition for smart management of the fishery system is the availability of a negative feedback loop based regulatory measure that can constantly work to reduce the difference between the actual and desired states of the system. Fishing feet overcapacity remains the fundamental problem of the Common Fisheries Policy (CFP) while the European fleets remain far too large for the resources available. It is believed that introducing the principles of Europe 2010 Strategy, EU IMP and of the MSP as a key instrument for the IMP implementation, the fishery property rights, e.g. the Individual Transferable Quota (ITQ) system, the Marine Stewardship Council's Principles and Criteria into the Baltic Sea fishery management would contribute to the actual removal of excess fishing capacity. It is suggested that the reversal of burden of proof, when and as necessary, could be technically performed based on the legal procedure of the Environmental Impact Assessment that gives the fishing industry the opportunity to demonstrate that it operates responsibly in return for access to fishing.

Keywords: Baltic Sea, governance of marine spaces, maritime spatial planning, fishery management, reversal of the burden of proof

* Contact author: Robert Aps: Estonian Marine Institute, University of Tartu, 14 Maealuse Str., 12618 Tallinn, Estonia [tel: +3725062597, fax: +3726718900, e-mail: robert.aps@ut.ee]

Introduction

Europe 2020: a strategy for smart, sustainable and inclusive growth" puts forward three mutually reinforcing priorities: 1) developing an economy based on knowledge and innovation, 2) promoting a more resource efficient, greener and more competitive economy, and 3) fostering a high-employment economy delivering social and territorial cohesion [1]. EU maritime affairs are addressed by the Integrated Maritime Policy for the European Union (IMP) [2] based on the clear recognition that all matters relating to Europe's oceans and seas are interlinked, and that sea-related policies must develop in a joined-up way if we are to reap the desired results.

The EU Marine Strategy Framework Directive (MSFD) [3] is seen as an environmental pillar of the IMP and it requires Member States to achieve good marine environmental status by 2020, to apply an ecosystem approach, and to ensure that pressure from human activities is compatible with good environmental status. Moreover, the Member States are required to cooperate where they share a marine region or sub-region and use existing regional structures for coordination proposes, including with third countries.

IMP considers the Maritime Spatial Planning as a fundamental tool for the sustainable development of marine areas and coastal regions, and for the restoration of Europe's seas to environmental health [4]. Although the decision-making competence in this area lies with the Member States the commitment at European level to common principles and guidelines is necessary to facilitate the process in a flexible manner and to ensure that regional marine ecosystems that transcend national maritime boundaries are respected.

While the implementation of MSP is responsibility of the Member States its overarching objective is to balance sectoral interests and achieve sustainable use of marine resources in line with the EU Sustainable Development Strategy by addressing the economic, social and environmental aspects on an equal footing [5].

In line with the IMP, the ecosystem approach is an overarching principle for MSP that operates within three dimensions, addressing activities on the sea bed, in the water column and on the surface while time is also taken into account as a fourth dimension [2]. EU IMP Action Plan [6] stipulates that integrated MSP across EU waters is a fundamental requirement for the continued sustainable development of maritime economic activities, because it provides a neutral tool to arbitrate between conflicting or competing activities or interests. It is clearly stated that MSP should be legally binding if it is to be effective and this might raise the issue of the appropriate administrative framework for MSP.

The EC Green Paper on Reform of the Common Fisheries Policy [7] states that 88 % of Community stocks are being fished beyond the Maximum Sustainable Yield (MSY) and 30 % of these stocks are outside safe biological limits, which means that they may not be able to replenish. Furthermore, despite continued efforts, fleet overcapacity remains the fundamental problem of the Common Fisheries Policy (CFP) while the European fleets remain far too large for the resources available and this imbalance is at the root of all problems related to low economic performance, weak enforcement and overexploited resources.This Chapter addresses some options for reduction of excess fishing capacity through more efficient and smart fishery management based on the EU Integrated Maritime Policy.

FROM PLANNING TO GOVERNANCE OF THE MARINE SPACE

The roots of the Europe 2020: strategy for smart, sustainable and inclusive growth, the Integrated Maritime Policy for the European Union and the EU Common Fisheries Policy (CFP) can be found in the European model of public management. Drechsler [8] argue that the EU paradigm for times to come in all of Europe is a state structure, and what is more, the EU is a Continental "state" organized and working along Continental French and/or German lines while the New Public Management (NPM) comes from Anglo-America, and it originates from the 1980s with their dominance of neo-liberal governments.

König [9] argues that the challenge faced today by continental Europeans in their confrontation with a NPM goes beyond the claim to an internal rationalization of the public administration by means of good management while the new diction is the language of the market, of competition, of enterprises, services, customers and of entrepreneurial management symbolizing the departure from the old administrative management. However, it is stated [8] that the contemporary European model of public management is most probably slowly thriving towards the hybrid Neo-Weberian State (NWS) model. For the EU CFP it means functioning "strong state" to back reinforcing the smart elements of the Baltic Sea fishery system, such as: 1) further regionalization to provide a more meaningful framework for coordinating different strands of marine policy - fisheries, environmental protection, maritime development, 2) development of the MSP as a main tool for managing fish stocks at Maximum Sustainable Yields (MSY) and providing a better future for the European fishing community, 3) further development of the property rights based fisheries management, and 4) the reversal of burden of proof if and as necessary.

It is stated [10] that the governance is the process of decision-making with a view to managing change as societies and organizations pursue their objectives. Sutherland and Nichols [11] argue that the good governance is based on recognition of the interests of all stakeholders and inclusion of their interests that can be expressed in a variety of ways: sovereignty, jurisdiction, administration, ownership (title), lease, license, permit, quota, customary rights, aboriginal rights, collective rights, community rights, littoral rights, public rights, rights of use, and public good.

Nichols et al. [12] suggests that the marine space governance should be linked to the law and information including 1) allocation of resource ownership, control, stewardship and use within society, 2) regulation of resources and resource use (e.g., environmental protection, development and exploitation, rights to economic and social benefits), 3) monitoring and enforcement of the various interests; adjudication of disputes, including inclusive processes, 4) management of spatial (geographically referenced) and other types of information to support all of the above functions.

According to Douvere and Ehler [13] „In its broadest sense, marine spatial management is about analyzing and allocating parts of three-dimensional marine spaces to specific uses, to achieve ecological, economic, and social objectives that are usually specified through the political process. Marine spatial management aims to provide a mechanism for a strategic and integrated plan-based approach for marine management that makes it possible to look at the „bigger picture" and to manage current and potentially conflicting uses to reduce the cumulative effects of human activities, and to deliver marine protection."

Pomeroy and Douvere [14] argue that the ultimate decision on what space will be allocated for what use (or non-use) is a matter of societal choice because the people are central to the decision-making process and are the agents for change.

It is stated [15] that fisheries management at both European and local levels is already strongly influenced by a spatial approach: 1) Total Allowable Catches (TACs) and quotas - the cornerstone of the CFP's stock conservation strategy – are set for International Council For the Exploration of the Sea (ICES) areas, 2) closed areas are used to protect commercial species at vulnerable stages in their life cycles, 3) multi-annual recovery plans for have attempted to deploy spatial controls through the designation of areas where fishing is permitted under particular conditions.

EC Green Paper on Reform of the Common Fisheries Policy [7] is addressing the integration of the Common Fisheries Policy into broader maritime policy context is stating that capture fisheries and aquaculture compete increasingly with other maritime sectors for marine space. Therefore, the marine spatial planning is an important element of the IMP with which the future CFP must be integrated. It seems to be quite natural to include the fishery management as an element among others into the framework of the emerging marine space governance.

FISHERY SYSTEM

Systems thinking that concerns an understanding of a system by examining the linkages and interactions between the elements that comprise the entirety of the system is used by Charles [16] when analyzing sustainable fishery systems and considering the following sub-systems of the fishery system: 1) the natural system (fish, ecosystem, biophysical environment), 2) the human system (fishers, post harvest sector and consumers, fishing households and communities, social/economic/cultural environment), and 3) the fishery management system (fishery policy and planning, fishery management, fishery development and fishery research).

Systems approach is usually applied to modeling of the complex fishery systems and/or their sub-systems [17-18]. In general, conception of the fisheries system as a set of interacting or interdependent entities forming an integrated whole is close to the conception of the coupled human and natural system [19].

Addressing the biological objectives of the fishery management Punt and Smith [20] argue that MSY has been, and to some extent remains, a key paradigm in fisheries management science that is representing the traditional key objective for fisheries management: removals should be as large as possible but nevertheless sustainable in long term. Commenting on MSY concept in general Larkin [21] argues that „The approach must be anthropocentric. It is a contradiction in terms to speak of biological objectives of fisheries management. Much more logical is to speak of biological constraints to management ... The real questions are: what should be the biological constraints and what should be the social objectives". Sissenwine and Symes [22] say that "Somewhat unwillingly, as a result of decisions made at the World Summit on Sustainable Development (WSSD), DG Fish has been obliged to adopt the concept of maximum sustainable yield (MSY) as a strategic goal for the management of Europe's fisheries." Nowadays MSY still remains a key concept in fisheries management emerging as a limit rather than as target reference point.

Economists traditionally use economic (resource) rents as a measure of the net economic benefits attributable to a natural resource. The economic objective of fishery management is to maximize the net economic benefits (sustainable rents) flowing from the fishery [23]. It is observed also that the Maximum Economic Yield (MEY) is generated at an effort level somewhat below MSY and therefore a policy aimed at achieving maximum resource rent would be ecologically friendlier than a policy targeting. It is suggested that efficiency/inefficiency of fisheries may be measured as the difference between maximum rents obtainable from the fisheries and the actual rents currently obtained.

According to Sissenwine and Symes [22] the economic objectives closely relate to concepts of efficiency, profitability and competitiveness in a subsidy free environment for the sector as a whole and for individual enterprises in the catching and processing areas while they must also concern the ability of the economic structures to evolve in line with the availability of natural resources, technological development and changing market opportunities. Christy and Scott [24] suggested already in 1965 that the 'maximize sustained yield' objective should be replaced by a 'maximize rent from the sea' objective.

The EC Green Paper on Reform of the Common Fisheries Policy [7] states that only a few EU fleets are profitable with no public support while most of Europe's fishing fleets are either running losses or returning low profits. World Bank and United Nations Food and Agriculture Organization (UN FAO) in their economic justification for fisheries reform [23] argue that "The notion that harvesters (fishers) have an exclusive, rather than a partial and conditional, right to the benefits from marine fisheries has tended to obscure the quest for increased [negative] social and economic benefits to society as a whole. This study shows that, in aggregate, the benefits to society as a whole are negative; that society underwrites the sector, through subsidies, by paying the costs of fisheries management and through depletion of capital (fish wealth)."

It is argued [25] that the most governments understand the problem of rent dissipation and it is common to see the management objectives that, not only include the conservation of fish stocks, but also improvement of the economic contribution of fisheries into the national economies.

Baltic Sea Fishery

In European seas, chronic overfishing continues to be a major problem, largely because the forces driving overexploitation have not been properly addressed [26]. Regarding the Baltic Sea fisheries, the International Baltic Sea Fishery Commission (IBSFC) was responsible for the management of shared Baltic Sea fishery resources in 1974 - 2005. Aps et al. [27] argue that given the political and economic pressure inherent in Baltic Sea fishery management the IBSFC Contracting Parties, as maximizers of economic returns, attempted to ensure that a greater common value was allocated to them. Therefore, they systematically set the TACs for all unit stocks more than the sustainable levels of exploitation while the science-based advice then served mainly as the starting point for "talking up quotas".

Aps and Lassen [28] show that for all fish stocks, TACs have been set systematically more than the scientific advice based on sustainable exploitation. Authors use the language of Eagle and Thompson [29] and interpret this as "decision overfishing" while the evidences of extensive underreporting of catches are interpreted as "implementation overfishing". It is argued that despite the measures that have been taken by IBSFC to combat the situation this still means that a management body was knowingly accepting a situation of overfishing. It is concluded that the combination of decision overfishing and implementation overfishing, and not the management measures per se, could be the reason for the failure of depleted stocks to recover [28].

Fishing fleet overcapacity is generally considered to be a major cause of persisting overfishing, because it creates a strong incentive to catch more than is allowed and it undermines the control and enforcement efficiency. Fleet overcapacity is believed to be the driving force behind illegal, unreported and unregulated (IUU) fishing because under the pressure of cuts in quota and increasing fishing costs, vessel owners and fishers feel forced to break the rules to ensure continued payments on the vessels as well as a decent living for themselves [30].

Elements of the Smart Management

Regionalization

There is growing consensus that the regionalized approach to European fishery management would take into account the particular nature of the regions in a smarter way. It is stated [22] that „ ... the need for a regionalized approach goes beyond that of rational fisheries management. It provides a more meaningful framework for coordinating different strands of marine policy - fisheries, environmental protection, maritime development and the common instrument of MSP. In short it is the only sensible design for developing IMM [Integrated Marine Management]." Authors suggest that the choice of institutional model for regionalizing the CFP lies between promoting the RACs to the status of regional management councils or tasking the relevant coastal EU Member States [e.g. Standing Conference of Ministers] to manage the region's fisheries on a collective basis using the RACs as their principal advisory partners. It is suggested also [22] that in any case „The European institutions would remain responsible for establishing the principles, broad objectives and strategic framework for sustainable development and for monitoring progress at the regional level. Responsibility for translating the principles etc into specific long term management plans would rest with the regional authority."

Maritime Spatial Planning

The MSP based fishery management is expected to be smarter than the recent command-and-control based management. MSP related decision-making competence lies with the Member States while the Common Fisheries Policy (CFP) is exclusive EU competence. IMP [1] states that managing fish stocks at Maximum Sustainable Yields (MSY) will provide a better future for the European fishing community and ensure its contribution to Europe's food security. IMP underlines that fisheries management must take more into account the welfare of coastal communities, the marine environment and the interaction of fishing with other activities. Methodology of MSP is seen as a suitable tool for integrated maritime policy development, and such a planning is a fundamental requirement for the continued sustainable development of maritime economic activities, including fisheries, because it provides a neutral tool to arbitrate between conflicting or competing activities or interests.

Coming age of Maritime Spatial Planning is bringing the Baltic fishing industry at the negotiation table with the representatives of other sea use interests (maritime transport, wind farms, mineral and oil extraction etc.) with aim to negotiate the allocation of the marine space concerned [31]. In order to meet the challenge the Baltic Sea RAC should develop sufficient knowledge base and argumentation setting to be efficient partner in a fully comprehensive, integrated, plan led system of management for the present and future exploitation and development of marine resources, including fisheries.

Symes [15] states that the fishing industry's concern at the prospect of MSP is rooted in its unique pattern of space use while fishing involves an extensive form of space use (laterally and vertically and endemic uncertainty demands flexibility in the use of space. Therefore, in order to increase the capacity of the Baltic Sea Region fishing industry for informed participation in a process of the MSP 1) the impact of actual or planned competing sea uses on spatial/temporal allocation of fishing possibilities for the Baltic fisheries should be assessed, and 2) the credible, relevant and sound arguments should be generated to be used by the stakeholders when balancing environmental, economic and social interests in a process of the Maritime Spatial Planning.

Property Rights Based Fishery Management

The substantial body of published evidences shows that with open-access, non-exclusive fishing rights, too many resources are concentrated into fishing. The Individual Transferable Quota (ITQ) system remarkably improves the allocation of resources because the fishing right-holders have a greater vested interest in the resource and are better motivated to take more responsibility for management. Scott [32] states that "There are three general powers of ownership: to manage the asset, to transfer or sell it, and to take the income from it. The ITQ certainly gave the fisherman the third power. Instead of merely allowing him to go out and compete with others until the TAC was reached, it entitled him to a definite fixed percentage of the TAC - that is to the yield itself." It is widely demonstrated that introduction of ITQ system is efficiently contributing to the actual removal of excess fishing capacity [32-33].

The Marine Stewardship Council's Certification

The Marine Stewardship Council (MSC), an international non-profit organization [34], has set an environmental standard to measure and reward well-managed fishery provided that 1) the fishing activity is at a level which is sustainable for the fish population, 2) fishing operations are managed to maintain the structure, productivity, function and diversity of the ecosystem on which the fishery depends, and 3) the fishery is meeting all local, national and international laws and has a management system in place that is able to respond to changing circumstances and maintain sustainability. It is expected that the MSC potential certification of the Baltic Sea fishery would mobilize the consumer community, the seafood market and the associated fishery to a more sustainable basis including substantial reduction of fishing fleet overcapacity.

Reversal of the Burden of Proof

With a wiev to reverse the current situation, among other measures, it is proposed [7] that the fishing industry should be given more responsibility through self-management: „Results based management could be a move in this direction: instead of establishing rules about how to fish, the rules focus on the outcome and the more detailed implementation decisions would be left to the industry. Public authorities would set the limits within which the industry must operate, such as a maximum catch or maximum by-catch of young fish, and then give industry the authority to develop the best solutions economically and technically. ... It would have to be linked to a reversal of the burden of proof: it would be up to the industry to demonstrate that it operates responsibly in return for access to fishing.“

The Precautionary Principle is usually defined as follows [35]: "When an activity raises threats to the environment or human health, precautionary measures should be taken, even if some cause-and-effect relationships are not fully established scientifically. In this context, the proponent of an activity, rather than the public, should bear the burden of proof [for the safety of the activity]." According to this definition, the Precautionary Approach clearly shifts the burden of proof: fishing is only permitted when it is proved that the activity will not cause undue harm to fish stocks or the marine ecosystem rather than the present situation when fishing restrictions are only accepted if the fishing activities are proved harmful.

In the context of the Baltic Sea MSP the fisheries is representing just one of the sea space use interests. Spatial allocation of the fishing grounds in the future most probably will be based on the different individual fishing rights. Technically speaking, one of the preconditions for the fishing grounds and/or fishing quota allocation could be the production of satisfactory Environmental Impact Assessment (EIA) for that particular fishery. Production of EIA in EU is sufficiently backed legally and can be used as evidence that the particular fishery operates responsibly in return for access to fishing. In addition, successful Marine Stewardship Council (MSC) certification of that fishery could be seen as an additional evidence of responsible fishing.

Conclusion

Baltic Sea fishery management is characterized by the regulatory overfishing: TACs are often set more than the scientific advice based on sustainable exploitation (decision overfishing) while the evidences of extensive underreporting of catches are interpreted as implementation overfishing. Consequently, many fishery resources are harvested neither at MSY nor at MEY and the resource rent is eventually dissipated.

It is believed that introducing the principles of EU IMP and of the MSP as a key instrument for the IMP implementation, the Fishery Property Rights (e.g. ITQ system), the MSC's Principles and Criteria into the Baltic Sea fishery management would contribute to the actual removal of excess fishing capacity. It is expected also that e.g. the ITQ system would create and enforce the missing smart negative feedback loop that will be constantly pushing the fishery system towards higher economic efficiency and ecological sustainability.

It is suggested that the reversal of burden of proof could be technically performed based on the legal EIA procedure that gives the fishing industry the opportunity to demonstrate that it operates responsibly in return for access to fishing.

REFERENCES

[1] EC. (2010). Europe 2020. A European strategy for smart, sustainable and inclusive growth. *COM(2010) 2020 final,* Brussels, 35 p.

[2] EC. (2007). An Integrated Maritime Policy for the European Union. *COM(2007) 575 final.* Brussels, 16 p.

[3] EC. (2008). Directive 2008/56/EC of the European Parliament and of the Council of 17 June 2008 establishing a framework for community action in the field of marine environmental policy (Marine Strategy Framework Directive). *Official Journal of the European Union,* L 164/19, 22 p.

[4] EC. (2008). Communication from the Commission: Roadmap for Maritime Spatial Planning: Achieving Common Principles in the EU. *COM(2008) 791 final*, Brussels, 12 p.

[5] EC. (2009). Mainstreaming sustainable development into EU policies: 2009 Review of the European Union Strategy for Sustainable Development. *COM(2009) 400 final.* Brussels, 16 p.

[6] EC. (2007). An Integrated Maritime Policy for the European Union: Action Plan. *SEC(2007)1278*, 38 p.

[7] EC. (2009). *Reform of the Common Fisheries Policy: Green Paper*. Brussels. 24 p.

[8] Drechsler, W. (2005). The Rise and Demise of the New Public Management. *Post-autistic economics review,* 33, pp. 17-28.

[9] König, K. (1997). Entrepreneurial management or executive administration: the perspective of classical public administration. In J.M. Kickert (Ed.). *Public management and administrative reform in Western Europe*, pp. 213-232. Cheltenham – Northampton, MA, Elgar.

[10] Sutherland, M.D. (2005). Marine Boundaries and Good Governance of Marine Spaces. *Ph.D. dissertation, Department of Geodesy and Geomatics Engineering, Technical Report No. 232,* University of New Brunswick, Fredericton, New Brunswick, Canada, 371 p.

[11] Sutherland, M.D & Nichols, S.(2006). Administering marine spaces: international issues. *FIG publication* 36, Copenhagen, 16 p.

[12] Nichols, S., Monahan, D. & Sutherland, M.D. (2000). Good Governance of Canada's Offshore and Coastal Zone: Towards and understanding of the Maritime Boundary Issues. *Geomatica,* Vol. 54, No. 4, pp. 415-424.

[13] Douvere, F., & Ehler, C. (2009). Ecosystem-based marine spatial management: an evolving paradigm for the management of coastal and marine places. In A. Chircop, S. Coffen-Smout & M. McConnell Eds. *Ocean Yearbook*, 23, pp. 1 – 26.

[14] Pomeroy, R., & Douvere, F. (2008). The engagement of stakeholders in the marine spatial planning process. *Marine Policy*, pp. 816– 822.

[15] Symes, D. (2005). Marine spatial planning: a fisheries perspective. *Report toEnglish Nature*. Unpublished, 35 p.

[16] Charles, A.T. (1994). Towards sustainability: the fishery experience. *Ecological Economics*, 11, pp. 201–211.

[17] Lane, D.E., & Stephenson, R.L. (1995). Fisheries management science: the framework to link biological, economic, and social objectives in fisheries management. *Aquat. Living Resour.,* 8, pp. 215-221.

[18] Garcia, S.M., & Charles, A.T. (2007). Fishery systems and linkages: from clockworks to soft watches. *ICES Journal of Marine Science,* 64, pp. 580–587.

[19] Liu, J., Dietz, T., Carpenter, S.R., Alberti, M., Folke, C., Moran, E., Pell, A.N., Deadman, P., Kratz, T., Lubchenco, J., Ostrom, E., Ouyang, Z., Provencher, W.,Redman, C.L., Schneider, S.H., & Taylor W.W. (2007). Review: Complexity of Coupled Human and Natural Systems. *Science,* Vol. 317, No. 844, pp. 1513 – 1516.

[20] Punt, A.E., & Smith, A.D.M. (2001). The gospel of maximum sustainable yield in fisheries management: birth, crucifixion and reincarnation. In D. Reinolds, G.M. Mace, K.H. Redford, & J.G. Robinson Eds. *Conservation of exploited species.* Cambridge University Press, Cambridge, pp. 41- 66.

[21] Larkin,P.A. (1988). Comments on the workshop presentations. In W. S. Wooster Ed. *Fishery Science and Management: Objectives and Limitations. Lecture Notes on Coastal and Estuarine Studies.*Vol 28, Springer-Verlag, New York, pp. 287 – 289.

[22] Sissenwine, M., & Symes D. (2007). Reflections on the Common Fisheries Policy. *Report to the General Directorate for Fisheries and Maritime Affairs of the European Commission.* 75 p.

[23] WB-FAO. (2008). The sunken billions: the economic justification for fisheries reform. *World Bank and Food and Agriculture Organization, Agriculture and Rural Development Department.* The World Bank. Washington DC. 86 p.

[24] Christy, F.T., & Scott, A. (1965). The common wealth in ocean fisheries: some problems of growth and economic allocation. Resources for the Future. Baltimore and London. *The Johns Hopkins Press.*

[25] [Shrank, W.E., Arnason, R., Hanneson, R. (2003). *The cost of fisheries management.* Ashgate Publishing. 302 p.

[26] EEA. (2005). Sustainable use and management of natural resources. *European Environmental Agency Report,* 9. 72 pp.

[27] Aps, R., Kell, L.T., Lassen, H., & Liiv, I. (2007). Negotiation framework for Baltic fisheries management: striking the balance of interest. *ICES Journal of Marine Science*, 64, pp. 858–861.

[28] Aps, R., & Lassen, H. (2010). Recovery of depleted Baltic Sea fish stocks: a review. *ICES Journal of Marine Science*, 67, pp. 1856–1860.

[29] Eagle, J., & Thompson, B.H. (2003). Answering Lord Perry's question: dissecting regulatory overfishing. *Ocean and Coastal Management,* 46, pp. 649–679.

[30] ORCA-EU. (2008). A report on IUU fishing of Baltic Sea Cod. N. Sporrong Ed. *Report commissioned from ORCA-EU.* The Fisheries Secretariat, Bromma, Sweden. 71 pp.

[31] Aps, R., Fetissov, M., Kell, L., & Lassen, H. (2009). Baltic Sea Regional Advisory Council as a hybrid management framework for sustainable fisheries. In C.A. Brebbia, & E. Tiezzi Eds.. *Ecosystems and Sustainable Development.* WIT Press, UK, pp. 163 - 172.

[32] Scott, A. (2000). Introducing property in fishery management. In R. Shotton Ed. *FAO Fisheries Technical Paper 404/1.* FAO, Rome, pp. 1-13.

[33] R. Arnason. (2002). A review of international experiences with ITQs: an annex to Future options for UK fish quota management. *CEMARE Rep. No. 58,* 64 p.

[34] MSC. (2010). *MSC fishery standard: principles & criteria and criteria for sustainable fishing*. Marine Stewardship Council, 8 p.

[35] Miller, N. 2009. *Environmental politics*. Routledge, NY, 207 p.

In: Fishery Management
Editor: Janice S. Intilli
ISBN: 978-1-61209-682-7

Chapter 7

FISHERIES MANAGEMENT IN NIGERIA: A CASE STUDY OF THE MARINE FISHERIES POLICY

F.M. Nwosu*[*], *E.O. Ita*[†] *and U.I. Enin
Institute of Oceanography, University of Calabar
PMB 1115, Calabar, Nigeria

ABSTRACT

We reviewed the marine fisheries laws as contained in the Sea Fisheries Decrees of Nigeria promulgated in 1971, 1972 and revised in 1992. We found out that these decrees were lacking in scientific data needed to effectively enforce the laws, both for industrial and artisanal fisheries. The major habitats of the different species, their breeding and nursery grounds; the migratory movements and seasons for different species; size selection of the fishing gears for different species caught simultaneously by relevant gears; sizes at first maturity of the different species; feeding/trophic relationships of the various species; maximum and optimum sustainable yields of the resource, and the corresponding effort, are some of the missing scientific information. This paper stresses the importance of these scientific data requirements for effective policy options and sustainable fisheries management.

1. INTRODUCTION

Nigeria has an extensive coastline of approximately 900 km and an Exclusive Economic Zone (EEZ) of about 217,313 km^2 (SeaAroundUs, 2007). Within the nation's EEZ, there is abundance of *Sardinella* and bonga and extensive breeding and feeding grounds for prawns. Because of the high export value of these prawns, many trawlers engage in shrimping with small mesh codend trawl nets less than the stipulated 44 mm, which expectedly exploit not only shrimps but juveniles of important commercial fish species stipulated for exploitation with 76mm trawl codend. In addition to fishing with illegal mesh sizes, the shrimp trawls also shrimp in shallow areas (less than 20 m depth or 5 nm strip) reserved for the artisanal

* E-mail address: fmnwosu@yahoo.com, francismnwosu@yahoo.com. Mobile: +234-803-835-5564
† Mr. E. O. Ita, a pioneer of fisheries management in Nigeria, passed on in April 2011.

fisheries by law. The by-catch from shrimping made up of small sized fishes were initially dumped in the sea, but because of the waste of biomass they were later permitted to be landed and sold to fish traders. This has created the danger of over- fishing of the important commercial fish species by shrimp trawlers. It is thus imperative that the existing Sea Fisheries Laws and Regulations be updated with necessary scientific data previously not available, in light of the non-sustainable exploitation of fisheries resources in the coastal waters. It is in this regard that the fisheries sector of the national economy can be rescued from imminent collapse. Moreover, government's poverty alleviation measures in the rural coastal communities would suffer a great set back because of the fisheries resources-dependent livelihood of the population. In addition, health problems related to deficiency of protein, which is most cheaply derived from fish diet, would attain significant dimension with increasing fish scarcity and non-affordability. This paper is targeted at highlighting the crucial scientific data needed for an effective monitoring and control of the operation of the fishing industry to enhance sustainability of the resources.

2. Problem Statement

In order to control marine fisheries exploitation, a number of Decrees were promulgated by government, starting form 1971, and additional regulations added as they became necessary all through to the latest Decree of 1992. In spite of these laws and regulations, fisheries exploitation is showing clear signs of severe overexploitation. This is largely because the existing laws and regulations lack fundamental scientific data on which sustainable exploitation depends. In particular, fin-fish selectivity by trawl nets is largely unknown.

It is well known that only about 5% of the total landings are attributed to industrial marine fisheries using trawlers, the remaining 95% being landed by local or artisanal fishers. Yet there is no provision in the Fisheries Decrees for operation of coastal canoe fishers (beside restriction to nearshore waters not beyond 5 nautical miles), even though this class of fisher folks operate in the most productive coastal waters where the probability for exploiting undersized fishes are highest. Information for stipulating gear types to be used within different coastal waters is also lacking. Consequently, the artisanal fishers exploit the choice commercial species like bonga, croakers, grunters, barracudas, mullets, sharks, rays, skates, snappers, soles, threadfins, cat fishes, shrimps, etc. without restriction. It is very clear that without proper scientific data the existing fisheries regulations will never halt the increasing trend of over exploitation of these living resources of our marine waters.

3. Critical Review of the Sea Fisheries Decrees

3.1. The 1971 and 1972 Sea Fisheries Decrees

As interim measures pending the collection and collation of scientific information, the Federal Government through the Supreme Military Council and using advice from Federal Department of Fisheries, promulgated the Sea Fisheries Decrees of 1971, 1972 and 1992. These decrees were limited in scope and were directed at the following parameters:

(i) Registration and licensing of fishing trawlers operating in the coastal waters of Nigeria. The number of trawlers to be registered required biological information on the Maximum Sustainable Yield (MSY) of the exploited stock and the yearly quota for each trawler. The catch quota for each boat was not included in the decree.
(ii) The 1971 decree prohibited the use of explosives and poisons in catching fish. The damages caused by explosives and poisons to the fish stock was known and did not require additional information prior to its promulgation.
(iii) The 1972 Sea Fisheries Decree restricted fishing trawlers from operating within the first two nautical miles of the continental shelf. This was intended to prevent industrial vessels from competing with small-scale artisanal canoe fishers. This restriction was later extended to 5 nautical miles in 1992. This decree was based on complaints, by local fishers, of damages done to their fishing nets by industrial fishing trawlers.
(iv) The 1972 decree also placed restriction on the codend of trawl nets used by industrial trawlers. The minimum codend mesh size of trawl nets was put at
(v) 3.5 inches (or 76mm) for finfish and 1.75 inches (or 44mm) for shrimping. This restriction was based on experience of trawling along the West African coast. It was lacking in details of the sizes of fish to be caught. Many undersized fishes were caught by shrimp trawlers initially and dumped in the sea for fear of being arrested. This resulted in the extension of the decree to include restriction on dumping of fish by-catch by shrimp trawlers.
(vi) The 1972 decree also restricted shrimp trawlers from operating within the inshore waters of the Lagos-West fishing grounds. This was intended to protect the juveniles of croakers common in the area. It was the only restricted fishing area for trawlers. There are many such areas along the Nigerian coast, particularly along the estuaries of major rivers, which constitute breeding and nursery grounds for different marine fish species.

These have to be identified and included in the revised law. As Neiland *et. al. (*2002) pointed out, weak scientific data lead to incomplete information and knowledge, and policies based on such data will remain incomplete.

Another crucial gap in Nigeria's fisheries policy is the complete absence of stakeholders' input during the policy formulation process. Experience shows that top-bottom approach to fisheries management does not yield the best of results. Local communities whose livelihoods derive from fisheries resources either directly or indirectly (e.g. artisanal fishers, women trading in fisheries products), local authorities where these resources domicile, commercial fishers, fishing equipment manufacturers and suppliers, should be given the chance to contribute to the policy formulation. Nigeria is now a democracy, so it is imperative to address these issues when the proposed revision of the laws takes effect.

3.2. The Exclusive Economic Zone (EEZ) Decree of 1978

By this law, Nigeria claimed 200 nautical miles as its EEZ, where it has sovereign rights over all natural resources, in concert with the provisions of the UN Convention on the Law of the Sea. The EEZ decree did not explicitly state the role that local research would play in the management and exploitation of the living resources, although that was a requirement under the Law of the Sea. The decree however provided for the policing of the EEZ by law enforcement agents.

3.3 The 1992 Sea Fisheries Decree

The 1992 Sea Fisheries Decree consists of two segments namely:

The Sea Fisheries Licensing Regulation of 1992 which stipulates the conditions for granting fishing licence for shrimping and fishing. This requires application for pre purchase assurance and submission of feasibility studies.

The 1992 decree also re-emphasized the provision of regulations dealing with restriction of trawling within the first 5 nautical miles of the continental shelf. Trawl codend mesh size restriction was re-emphasized together with:

(i) Dumping of edible and marketable sea products and export thereof. Accordingly, fish landed by shrimp trawlers must not be less than 75% by weight of the total landing including the head on weight of the shrimp landed. This decree opens itself to wrong interpretation and has possibly resulted in the landing of more by-catch than shrimps. The rendering: "Fish landed by shrimp trawlers should not be less than 75% by weight..." This implies that the by-catch (fish) should not be more than 25% of the total catch.

(ii) Restriction on minimum sizes of fish to be caught was not available at the time. The decree did not supply any information on fish sizes but made provision for the Nigerian Institute for Oceanography and Marine Research (NIOMR), Lagos, to publish the sizes of different species (minimum total length of different species) to be caught during each year.

The provision of such information required special commissioned surveys by the Development Agency (Federal Department of Fisheries) on account of the cost-intensive nature of such surveys. In the absence of such commissioned study, the gap in knowledge exists up to the present, despite the fact that the decree has stipulated for such information.

The fixing of 3.5 inch codend was based only on the selectivity of croakers, the most dominant species at the time and lacked scientific information on the sizes of other species likely to be caught by the 3.5 inch codend of trawls.

Because of this deficiency in the decree and the lucrative trade in the by-catch of small cheap small fish, some shrimp trawlers became more interested in the by-catches than shrimps meant for export. Such action was dangerous to the survival and continuous recruitment of the fish stocks.

3.4. Provision for Future Regulations by the Minister in 1992 Sea Fisheries Decree

Both the 1972 and 1992 decrees made provision for future regulations on:

(i) Areas to be prohibited for fishing, besides the Lagos-West fishing grounds.
(ii) Sizes of nets or mesh sizes to be employed in fishing and sizes of fish to be captured.
(iii) Any other regulation relating to the conservation and protection of the fish stocks in the sea.

In spite of the obvious deficiencies in the sea Fisheries Decrees so far promulgated, no attempt has been made by the Development Agency to commission studies with the aim of updating the decrees with the relevant scientific information. There is obvious decline in fish landings by trawlers as a result of over exploitation of juvenile fish both by industrial fishers and coastal artisanal canoe fishers . Shrimping is being sustained because of the lucrative trade in the by- catch from shrimp trawlers. There is the need therefore to commission specific investigations to supply information for updating the 1992 decree and for additional regulations to control indiscriminate exploitation by coastal artisanal fishers.

3.5. Fisheries Regulation for Coastal Artisanal Fishers

The coastal artisanal fishers exploit the bulk of the fish from the coastal waters of Nigeria. Only about 5% of the total landings are attributed to industrial marine fisheries using trawlers. The canoe fishers exploit both marine and brackish water species common along the estuaries and lagoons. The limit of 5 nautical miles made exclusive to canoe fishers by virtue of the Sea Fisheries Decree, constitute the most lucrative fishing ground, being the shallowest portion of the continental shelf.

The canoe fishers penetrate all the breeding and nursery grounds and capture immature juveniles and breeding populations. The fishing gears used by canoe fishers are dangerous and include both active and passive gears such as cast nets, beach seine nets, purse seine nets, boat seines, gill-nets, traps, hooks etc.

The law made provision for additional scientific information on the total lengths of fish to be landed by both the industrial trawlers and coastal artisanal fishing gears. From all indications no such information has been collected and supplied to the Development Agency (the Federal Department of Fisheries) for updating the existing decrees.

Fisheries scientists from different higher institutions and research institutes have carried out biological studies of some marine fishes and published their findings but these studies were not directed towards the needs of the Development Agency for updating the existing laws and regulations. Such studies require special commission.

It is necessary to re-emphasize the need for such studies with the aim of filling the gap in existing knowledge for updating the Sea Fisheries Decrees so far promulgated.

4. The Missing Scientific Information

The present low level of the resources calls for urgent commissioning of studies aimed at collation of existing information on the resource and conducting new studies aimed at updating and collection of additional information. Such information will include:

(i) The major habitats of the different species and their breeding and nursery grounds. This information is needed for regulation on closed area and season to protect the breeding population and the juveniles.
(ii) The migratory movements and seasons for the different species also require investigation.
(iii) Catch composition of the various fishing gears used by both industrial and artisanal canoe fishers and the size selection of the gears for different species caught simultaneously by different fishing gears.
(iv) Minimum sizes of the different species at first maturity. This will be used in identifying the appropriate mesh size to be permitted for fishing to ensure that each fish reproduces at least once before they are captured.
(v) Feeding habits of the various species need be studied for comprehensive trophic level classification. The choice of minimum mesh size will depend on the species selected for protection. Usually, herbivorous species or primary consumers are the most popular species because of their abundance in the population followed by secondary consumers. The carnivores or tertiary consumers are often the largest in size and feed on the herbivores. Whereas the rich citizens prefer the bigger carnivorous species though expensive, the poor masses prefer the cheap smaller herbivores. Often, the minimum size of gear for catching some mature herbivores and secondary feeders may end up catching the majority of immature carnivores.

The ultimate goal is to utilize the information collected to draft a comprehensive Marine Fisheries Laws and Regulations covering both industrial and canoe fisheries. This will include methods of implementation to make the laws effective.

5. The Current State of Knowledge

5.1. Estimates of Potential Yield

Estimates of potential yield for Nigerian inshore and brackish water resources have been documented from the early sixties and updated by different workers up to the early nineties. Among these are Longhurst (1961) who estimated the potential yields of Nigerian marine resources to be about 40,000 metric tons. Ajayi (1982) estimated the potential yield for the Nigerian inshore waters (0 — 50 meters) to be 15,000 m.t. Also Ajayi & Talabi (1984) re-estimated the yield potential in relation to trawl codend mesh size. They concluded that if 52 mm (2 inch) codend is used the potential would range between 9,048 and 16,965 m.t. However if the legal codend mesh size of 76mm (3.5 inch) was used, a potential of about 20,000 m.t. is possible plus about 7500 m.t. of by-catch from shrimpers. This would bring the total potential yield to about 27,500 m.t. Enin (1997) estimated the maximum sustainable yield of the industrial shrimp fishery at 3447 m.t (Schaefer model) and 2706 m.t. (Fox model); and 20,000 m.t. for the industrial fin-fish fishery.

Ajayi and Talabi (op.cit) concluded that judging from than current estimates of landings of inshore industrial trawlers estimated at 16,342 m.t., the fishery was at the time of the estimate being exploited at about 60% of its potential. Elliot (1993) summarized the Nigerian coastal fisheries resources between 1980 and 1989 and observed a consistent increase in the number of trawlers registered and the estimated total landings based on statistics collected by the Federal Department of Fisheries (Table 1).

Table 1. Fish catch (metric tons) from inshore fishing and shrimping between 1980 — 1989

Year	Total Landings	No. of Trawlers	Average Landing
1980	13,631	80	170.4
1981	9,611	81	118.7
1982	18,861	86	219.3
1983	19,245	120	160.4
1984	25,650	133	192.9
1985	26,142	163	160.4
1986	25,042	250	100.2
1987	24,900	252	98.8
1988	35,608	372	95.7
1989	33,645	440	76.5

Source: Elliot (1993)

From the table above, the potential yield of 27,500 m.t. of Ajayi and Talabi (op.cit) was almost reached in 1985 with increase in the number of trawlers from 133 in 1984 to 163 in 1985 and with total landings of 26,142 m.t. That notwithstanding, the registration of more trawlers increased consistently to about 440 in 1989 with total landings for that year estimated at 33,645 m.t. but with the lowest average landing per trawler of 76.5 m.t. compared with 192.9 m.t. in 1984 when only 133 trawlers were registered.

5.2. Law Enforcement

Amiengheme (1993) outlined some of the constraints to the enforcement of fisheries laws and regulations and indicated lack of strong political commitment on the part of government as one of the major constraints. This implies that pressure was being put on the licensing agents to continue licensing even above the estimated sustainable yield of the fishery. There is need for coordination between the licensing authority of The Federal Department of Fisheries and the Nigerian Institute for Oceanography and Marine Research (NIOMR) in aspects of enforcement of number of boats to be registered.

Tobor (1991) states that the primary objective of licensing vessels was to generate revenue while the control of effort was secondary and ineffective. The licensing fee of N4,000 (Four thousand naira) was too low at the time to discourage investors. Even a hundred fold increase in licensing fee could not have discouraged interested investors.

5.3. Mesh Size Regulation

The stipulation of "two different mesh sizes: fin fish (76 mm or 3.5 inch) and shrimp (44 mm or 1.75 inch)"

is another controversial issue. The two vessels are all allowed to fish beyond 5 nautical miles from the coast line. Any fish that escapes from the 76mm codend trawl net is retained by the 44mm codend trawl net of shrimpers and designated as by-catch. Initially, the by-catch was discarded at sea but was later prohibited on account of wastage of biomass. Overfishing cannot be controlled with two different codend mesh sizes fishing along the inshore waters of Nigeria.

Ansa-Emmin (1980) in a discussion of the fisheries resources of Eastern Central Atlantic, states that the sixth session of the Committee for East Central Atlantic Fisheries (CECAF) recommended the adoption of a mesh with 60mm for the exploitation of all demersal species (shrimps, fin-fish and cephalopods). The research conducted in Nigeria and the Congo by CECAF showed that the existing mesh size of shrimp trawlers can be increased up to 60 or 70mm without significant reduction in shrimp landings while at the same time reducing juvenile mortality among other species of fish captured and discarded. Ghana adopted this recommendation and reduced the codend mesh size from 77mm to 60mm (Dowuona 1984). Nigeria is yet to adopt the CECAF recommendation.

5.4. By-Catch from Shrimping

Okpanefe (1982) sampled the by-catch from shrimp trawlers in 1981 and concluded that only 8.27% of fish by-catch was small fishes of 0-20cm while 91.25% were medium sized fishes of 21.0cm to 41.0cm. However in 1986 with increase in the number of shrimpers from less than 40 to over 100, Kusemiju (1993) reported that a large proportion of fish landed by shrimp trawlers were small-sized fishes. He recommended the setting up of a commission for the conservation and control of the fishery as is usually done in developed countries. This recommendation was very timely and should be implemented.

5.5. Coastal Artisanal Fishery

The potential yield from this sector which is estimated to land over 90% of total fish yield from the coastal industrial and artisanal fisheries has also been estimated. Using the estimated total number of coastal artisanal fishers in Tobor et al (1977), Ajayi and Talabi (1984) estimated the yield from coastal and brackish water artisanal fisheries at between 128,000 and 170,000 tons. Estimates for this sector vary according to different authors. Ssentongo et al. (1983) estimated 100,000 tons as the ceiling for inshore artisanal and industrial fisheries. If 95% of the catch is landed by the artisanal sector, then about 95,000 tons is attributed to this sector by the authors.

Judging from the Federal Department of Fisheries statistics (1971 to 1983), landings from coastal and brackish water fisheries varied from 199, 639 tons in 1971 to 376, 943 tons in 1983. This shows that this sector more than tripled Sentongo *et al* (op. cit) estimate of 95,000 tons. If the figure then was more than triple the potential, the current state of the fishery of this sector can be predicted. The yield is bound to decline as evidence of overfishing and the catch is expected to be dominated by small fishes. This sector exploits all the species captured by industrial vessels and in addition land substantial quantities of some pelagic species such as the bonga (*Ethmalosa fimbriata*) and *Sardinella* sp.

Holzlohner et. al. (2004) lists the fishing gears used by this group of fishers as driftnets with meshes ranging from 1½ to 4½ inches (34 — 115mm); set-nets with meshes of 54 — 113mm; cast nets with meshes of 10 —50mm. In addition there are purse seines for exploitation of the bonga with meshes of 50mm and below. Ama-Abasi and Holzloehner (2002) observed that in the outer estuary, the post-larval stages measuring 2-9cm are the target of boat seine fishery while the adult stages (18-25cm) are exploited by purse seine fishery It is important to ensure that the coastal artisanal fishers are taken care of in the Sea Fisheries Decree since they exploit the most productive sector of coastal waters.

6. Management Measures

The coastal waters of Nigeria can rightly be classified as an ecological disaster zone in view of the numerous problems affecting the zone, including pollution from industrial activities and oil exploration and overfishing by both industrial and artisanal fishers. It is relevant to set up a Commission within the cover of Natural Resources Conservation Council (NARESCON) to proffer solution for the conservation and control of the fishery resources to obviate a total collapse of the fishery.

The commission would be empowered to review the existing conservation measures of the coastal zone with special reference to fisheries and commission studies to update the existing policies. The major problem of policy formulation and implementation is availability of financial resources to meet specific needs such as:

a) Research to update the laws with relevant information as earlier listed.
b) Vessels for policing the coastal zone to enforce existing laws including the EEZ.
c) Updating of the national Fishery Statistics for estimation of total landings including types of gears and sizes of fish landed, and number of boats and fishers exploiting the fishery.

Experience in the past has shown that budgetary allocations to Fishery Departments and Research Institutes are too meager to meet the requirements of the above needs. Policy options and policy formation have political implications and should therefore be handled and financed through a Commission as recommended. Similarly, policy implementation also has some political implications which require enforcement by laws. Stipulation of number of trawlers to be registered each year cannot be enforced by an individual without a supporting law on account of political pressure as earlier noted.

7. Conclusion

The Sea Fisheries Decrees so far promulgated from 1971 to date require additional scientific information for effective enforcement. The need for commissioning of studies to fill the missing gaps in the decrees has been emphasized. The 1992 decree in particular noted the deficiencies in the degrees and therefore made provision for future regulations on.

(i) Areas to be prohibited for fishing along the coast.
(ii) Sizes of nets or mesh sizes to be employed in fishing and sizes of fish to be captured and
(iii) Any other regulation relating to the conservation and protection of the fish stocks in the sea.

The above provision took cognizance of the need to include coastal artisanal fisheries in future laws and regulations considering the fact that these fishers account for about 95% of total fish landed along the coast.

The poor state of the resource has been highlighted based on available information. Management measures recommended to remedy the deplorable state of the resource include:

- The setting up of a Commission under to Natural Resources Conservation Council with adequate financial resources to commission studies needed to update the laws and render them enforceable.
- The Commission is also expected to finance the provision of materials and equipment needed for policing the coastal zone and for updating the national fishery statistics which has been long outdated.

It is noted that policy options and policy formation mechanisms have political undertone and should be supported by an established Corn mission.

REFERENCES

Ajayi, T. 0. (1982), The maximum sustainable yields of the inshore fish and shrimp resources of the Nigeria continental shelf. *J. Fish. Biol.* 20: 571 — 577.

Ajayi, T. 0. and Talabi, S. 0. (1984). The potential and strategies for optimum utilization of the fisheries resources of Nigeria. *N.I.0.M.R. Technical Paper* No. 18, 25p.

Ama-Abasi, D. E. and S. Holzloehner (2002). Length frequency distribution of *Ethmalosa fimbriata* (Bowdich) in Çross River Estuary and adjacent coastal waters. *Global Journal of Pure and Applied Sciences* 8, 3: 299-304

Amiengheme, P. (1993). The role of Federal Department of Fisheries. In: *Proceedings of the National Conference on Conservation of Aquatic Resources*. Sponsored by Natural. Resources Conservation Council (NARESCON)

Ansa-Emmin, M. (1980). Fisheries Resources of the East Central Atlantic. In:*Proceedings of the National Seminar on Integrated Development of Artisanal and Inshore Fisheries in Nigeria*. Sponsored by Federal Department of Fisheries.

Dowuona, V. N. (1984), F. A. 0. *Project for the development of Fisheries in the Eastern Central Atlantic CAF/TECH/84/61*. Measures taken to regulate fishing capacity and effort in Ghana.

Elliot, O. (1993). Nigerian Coastal Fisheries Resources; In: *Proceedings of the National Conference on Conservation of Aquatic Resources.*Sponsored by Natural Resources Conservation Council (NARESCON).

Enin, U. I. (1997). The overfishing of Nigeria's inshore demersal fishery resources. *Transactions of Nigerian Society for Biological Conservation* 6: 1-5.

Holzloehner, S., F. M. Nwosu and U. I. Enin (2004). Fishing effort statistics of the artisanal fisheries of the Cross River Estuary, Nigeria. *Global Journal of Pure and Applied Sciences*, Vol. 10, No.2, March 2004: 249-256.

Ita, E. O. 1985 (edited). Transactions of the Inland Fisheries Laws and Regulations for Nigeria: *Verbatim report of proceedings of a round table conference, 1985*, published by the National Institute for Fresh Water Fisheries Research, New Bussa.

Ita, E. O. (1985). Guidelines for the drafting of Inland Fisheries Laws and Regulations for Nigeria. In *Transactions of the Inland Fisheries Laws & Regulations* (E. O. Ita edited)

Ita, E. O. (1993), Inland Fishery Resources of Nigeria. *CIFA (FAO) Technical Paper* No. 20 FAQ, Rome, Italy. 123p.

Kusemiju, K. (1993). Development, utilization and conservation of aquatic resources in Nigeria. In: Pro*ceedings of the National conference on conservation of aquatic resources.* Sponsored by N Resources Conservation Council (NARESCON).

Okpanefe, M. O. (1982). Sampling the catch of fishing trawlers. *NIOMR Technical Paper* No. 2. 25p.

Longhurst, A. R. (1961). *Report on the Fisheries of Nigeria.* Ministry of Economic Development, Lagos.

Neiland , A.E., Béné, C., Jolley, T., Ladua, B.M.B., Ovie, S., Sule, O., Baba, M., Belal, E.,

Tiotsop, F., Mindjimb, K., Dara, L., Zakara, A., and Quensiere, J. (2002). The Lake Chad Basin Fisheries: A strategic analysis of key policy issues affecting aquatic resource management, stakeholder livelihoods and economic development. In Neiland, A.E. and Béné, C. (eds) *Sustainable Development of African continental Fisheries: A regional study of policy-formation mechanisms in the Lake Chad Basin.* Final Report. Commission INCO Project No. ERBICI18CT980331.

SeaAroundUs (2007). A global database on marine fisheries and ecosystems. www.seaaroundus.org.

Sea Fisheries Decree of 1971. *Federal Government Gazette* Vol. 58 of 1871 Decree No. 30.

Sea Fisheries Decree of 1972. *Federal Government Gazette* Vol. 59 of 1972 Decree No. 30.

Sea Fisheries Decree of 1992. Federal Government Gazette Vol. 79 Decree No. 108.

Ssentongo, G. W., Ajayi, T. O. and Ukpe, E. T. (1983) *Report on a resource appraisal of the artisanal and inshore fisheries of Nigeria.* FAO/UNDP: FNDP/NIR/77/007 43P

Tobor, J. G. (1991) Marine fish resources of Western Africa: Potentials, Management, Development and Constraints to their utilization to satisfy increasing demand. *NIOMR Technical paper* No. 70. 25p.

In: Fishery Management
Editor: Janice S. Intilli

ISBN: 978-1-61209-682-7

Chapter 8

THE ROLE OF FISHERY-INDEPENDENT DATA

Richard E Thorne and Gary L Thomas
Prince William Sound Science Center, Cordova, AK
Rosenstiel School of Marine and Atmospheric Sciences, Miami, Florida

ABSTRACT

In 1998, the National Research Council committee on Fish Stock Assessment documented that most fisheries management is based on fishery-dependent forecasting methods. Recent studies have graded the quality of management programs by the amount of fishery-independent data they have on the stock size. We use two applications in Alaska to illustrate contrasting fishery-independent approaches: (1) a direct and precise estimate of the biomass of adult herring, and (2) an indirect study of the conditions in a pink salmon nursery area that could influence survival of the fry to adults. The first, an 18-year study of the herring stock in Prince William Sound, shows that fishery-dependent models twice failed to detect population collapses, and had lag-times of up to five years to detect major population changes. The second, a similar long-term study of the pink salmon stock in Prince William Sound, shows that the adult returns were a function of at least four dynamic biological variables during the spring fry outmigration. In the case of the Pacific herring, we document that the assessment by traditional fishery-dependent models resulted in measures of abundance that were too late, too unrepresentative or too imprecise to make correct management decisions. The failure for timely detection led to overfishing that accelerated a population collapse. The pink salmon example illustrates the case for true ecosystem-based fisheries management. The biological forcing mechanisms that caused change in pink salmon returns were neither understood nor measured. Fisheries-independent data and targeted environmental monitoring offer the best solution to improving the conservation of exploited fish stocks. This approach is a prerequisite for quality fishery management, but must include timeliness, robustness and precision.

INTRODUCTION

Concern for over-harvesting of marine fish stocks dates back to the late 1890s (Peterson 1894). Over the past decade, several prominent authors have warned of widespread, over-exploited fish populations (Casey and Worm 1998; Meyers and Worm 2003) and the debate has become contentious (Walters 2003; Hilborn 2006; Pauley. 2009). Recent controversy has centered on weaknesses of the data and methods and called for a return to "testable hypotheses and evidence" to determine the status of fish populations. The National Research Council (NRC) committee on Fish Stock Assessment documented that most fisheries management today is based on the forecasts of stock biomass by theoretical models and fishery catch data (Anon. 1998; Quinn and Deriso 1999; Quinn 2003), which are referred to as fishery-dependent methods. Fishery catch data include fishing vessel catch per unit effort and biological characteristics such as age composition, as well as the catch amount. The committee confirmed what many experts have recommended for decades, that every management program needs to develop fishery-independent monitoring of the stock size (Gunderson 1993). Fishery-independent methods include direct assessment using a variety of techniques such as research trawls or hydroacoustics. The NRC committee was nonspecific about which fishery-independent methods to use or the need for high quality data. In a more recent publication, marine fisheries with fishery-independent data were compared, but these data were deterministic and there was no call for the stochastic data needed to p[illegible] estimates are representative of the true stock size (Worm et al. 2009).

There have been several publications that discussed the errors associated with the measurement of fish density and the estimation of stock size (Gunderson 1993; Thomas et al. 1997; Thomas and Thorne 2003). Historic failures to make precise estimates of a stock size have undoubtedly discouraged development of direct empirical approaches and fostered the use of theoretical models (Thomas 1992). Nevertheless, large-scale ocean surveys are conducted routinely by government agencies using various midwater and bottom trawls and/or hydroacoustics (Gunderson 1993; Simmonds and MacLennan 2005). Unfortunately, most of these ocean surveys are not repeated and it is not known if the estimates represent the true stock size. In the only known comparison of forecast errors and fishery-independent surveys, the uncertainty in the forecasts exceeded 6-fold (Thorne and Thomas 2008). In the last decade there have been only a few repeated ocean surveys that have reproduced consistent estimates of stock size and these have documented the unreliability of a single survey to represent a fish stock (Vilhjalmsson and Carscadden 2002).

In this chapter, we use two case histories in Alaska to contrast fishery-independent methods with fishery-dependent approaches. First, an 18-year study of the herring stock in Prince William Sound (PWS) shows that fishery-dependent models twice failed to detect population collapses and had lag-times of up to five years to detect major population changes. The second case history is a long-term study of the pink salmon stock in Prince William Sound that showed the adult returns were a function of at least four dynamic biological variables during the spring fry outmigration. The fishery-independent techniques in this case are basic observations of ecological forcing functions that govern fish production and survival. Such an approach is seldom used in fishery management, but illustrates that good fisheries management ultimately needs to understand the ecological forcing functions that govern fish production.

CASE HISTORY 1- A DIRECT MEASUREMENT OF ADULT BIOMASS

Prince William Sound is a complex fjord/estuary (Thomas et al. 1991) located at the northern margin of the Gulf of Alaska. PWS supports a diverse assemblage of fish and wildlife, and Pacific herring *Clupea pallasi* is one of the most abundant forage fish available for piscivorous fishes and wildlife. The Prince William Sound herring stock collapsed after the 1989 *Exxon Valdez* Oil Spill (EVOS) (Peterson et al. 2007). The collapse motivated an annual series of direct observations on the abundance of the herring in 1993 that continues to date. Thorne and Thomas (2008) describe the data through 2005 and contrast the fishery-independent and fishery-dependent data sets. Herein we review and update these data. Our objective is to contrast the performance of fishery-dependent and independent methods, and to specifically investigate why the traditional management methods, primarily based on fishery-dependent data, did not detect the collapse that followed the oil spill. The multi-decadal data sets we examine include our own hydroacoustic estimates of adult herring biomass (1993-2009) and a mile-days index of abundance, which is the cumulative miles of spawn (milt) observed along the beaches during the spawning season (1973 to 2009). These two data sets are independent of fisheries data. Neither were used in fishery management until recent years. Instead, fishery management forecasts were based on aerial estimates of fish biomass from 1974 to 1987, estimates from egg deposition measurements from 1988-1992 (Schweigert and Stocker 1988; Becker and Biggs 1992; Brown 2007) and age-structured assessment (ASA) models since 1993 (Quinn et al. 2001; Hulson et al. 2008). The ASA model is primarily fishery-dependent. We document two major changes in the herring population since 1989. In both cases, there was a multi-year lag time before the management forecasts detected changes in the stock that required a management response. The failures of timely detection resulted in fish stock collapses. We argue that these lag times are inherent in the fisheries models that are used to manage most fisheries today, and these lag times are at least partially responsible for stock collapse. We show that fisheries-independent data collected prior to the fishery eliminate the lag times, offer the best solution to improving the conservation of exploited fish stocks and provide the quantitative baseline to evaluate environmental impacts. However, fishery-independent methods need to be precise to be useful for these purposes.

Methods

The hydroacoustic methods we use in PWS are detailed in several publications, including Thomas et al. (1997), Thomas and Thorne (2001) and Thorne and Thomas (2008). The general use of hydroacoustic methods in fisheries assessment are described in Simmonds and MacLennan (2005). Historical information on the abundance of Pacific herring in PWS includes aerial survey estimates of the length of spawn (milt) patches along beaches and age-structured analysis (ASA) model estimates. The Alaska Department of Fish and Game (ADF&G) has conducted the annual aerial surveys of spawn since 1973, so this is the longest and most consistent index of herring abundance. The ASA model has been run to forecast the PWS adult herring biomass most years since 1993, and has included several versions (Quinn 2001; Hulson et al. 2008). All versions of the model reconstruct the population history of the herring since 1980. Most versions show similar trends. However, revised estimates from versions of the ASA model are often substantially different from initial forecasts, particularly for 1993 and 1998-99.

Thomas and Thorne (2003) showed that the cumulative mile-days of milt index from the aerial surveys correlated well with the acoustic estimates from 1993 to 2002. Thorne and

Thomas (2008) updated the comparison among the acoustic estimates, mile days of milt index and ASA model estimates to 2005.

Results

The index of the miles of herring spawning along beaches, or the mile-days index, is the longest continuous data set. This index was built upon the observations of spotter pilots during the spawning season (the milt from the male herring creates a milky plume along the shorelines). The 36-yr database shows a general increase in the herring stock biomass from 1973 to 1988, followed by a stock collapse starting in 1989, the year of the *Exxon Valdez* oil spill (Fig. 1).

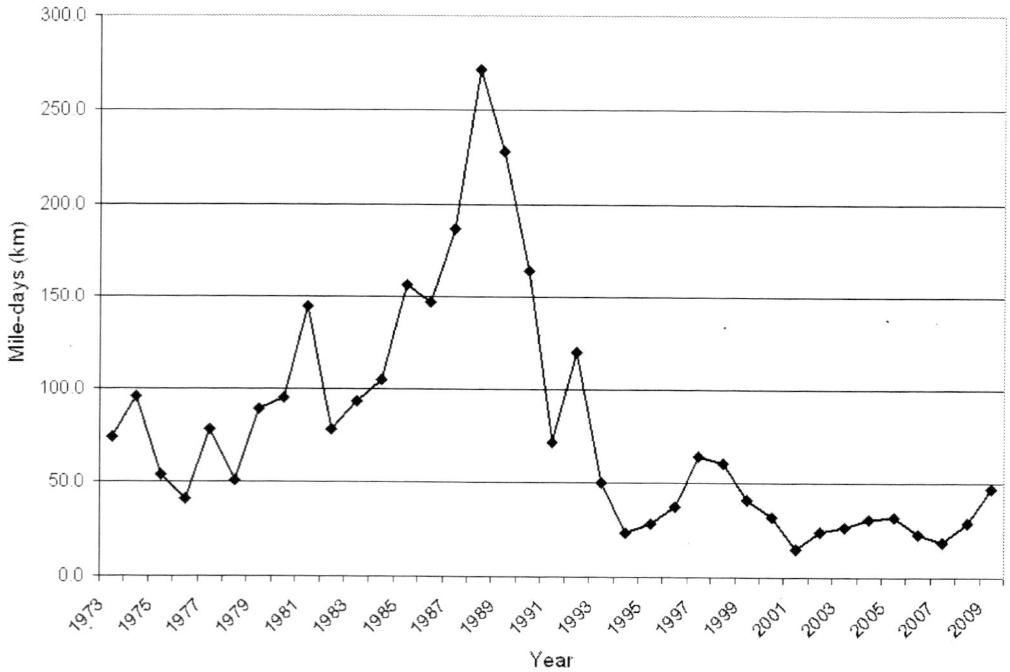

Figure 1. The 36-year database of the mile-days of spawn index.

Hydroacoustic surveys were initiated in fall 1993 after fishermen could not find fishable concentrations despite a management forecast for high abundance. The first estimate confirmed that the stock was seven times lower than the management forecast, or 18,812 instead of 133,852 t. The subsequent annual hydroacoustic estimates of herring biomass are correlated positively with the mile-days index (Thorne and Thomas 2008). In 2003, we converted the mile-days index to a mile-days biomass model using a regression between hydroacoustic estimates of biomass on the mile-days index (Thomas and Thorne 2003). This conversion allows a direct comparison of the mile-days estimates with the other estimates of the herring biomass. The mile-days biomass database agrees with the hydroacoustic measurements and/or the direction of change in most cases (Fig. 2).

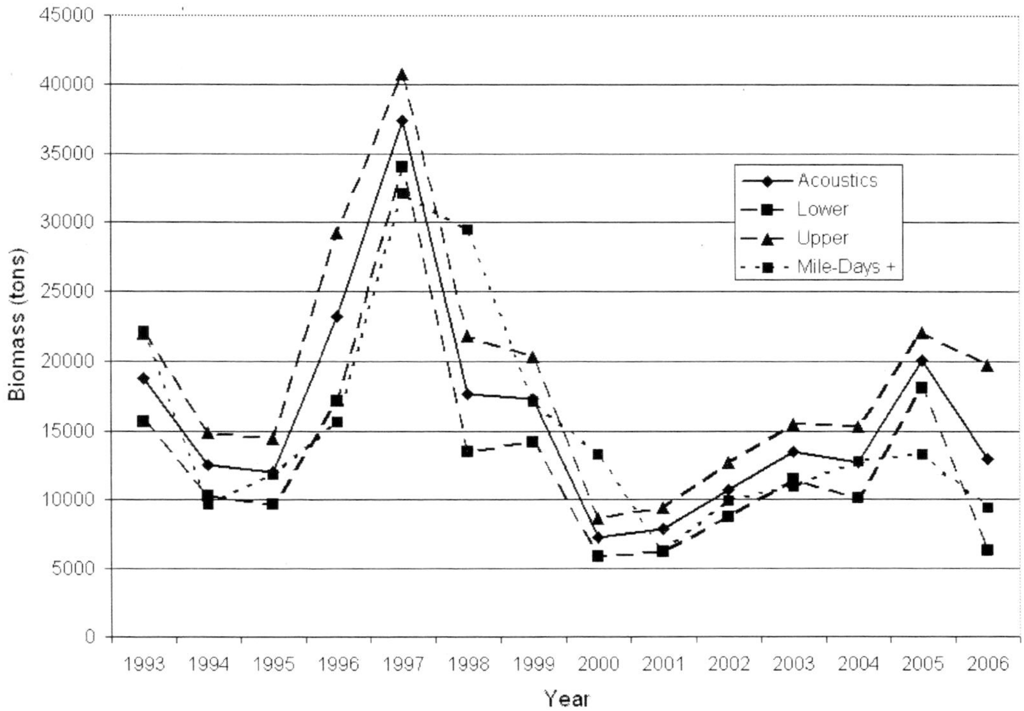

Figure 2. Comparison of the mile-days model and the hydroacoustic estimates with upper and lower 95% confidence intervals. Harvest values have been added to the mile-days estimates for standardization.

The fishery management data substantially lag the pattern shown by the mile-days estimates (Fig. 3). In particular, while the mile-days estimates drop precipitously after the 1989 oil spill, the management data based on egg deposition measurements in 1989-92 and the ASA model in 1993 portray a continuing high population. Consequently, fishery harvests during this time were the highest since the days of the reduction fishery (Brown 2007) with 65,000 tons removed.

The herring population began to recover after a low in 1994. By 1997 estimates from the hydroacoustics, the mile-days model and the ASA model forecast all agreed that the stock had recovered to a biomass of more than 30,000 mt, and a commercial fishery was reopened. However, after the resumption of the commercial fishery in 1997, both the hydroacoustic and mile-days estimates declined in 1998. In the case of the hydroacoustics, the decline was substantial. By 1998, both the hydroacoustics and mile-days estimates showed a substantial decline, but the ASA model-based forecasts did not decline until 2000 (Fig. 4). In 1989, the management forecast model moved in the wrong direction and did not detect the decline until five years later. For the 1998 decline the lag time was two years.

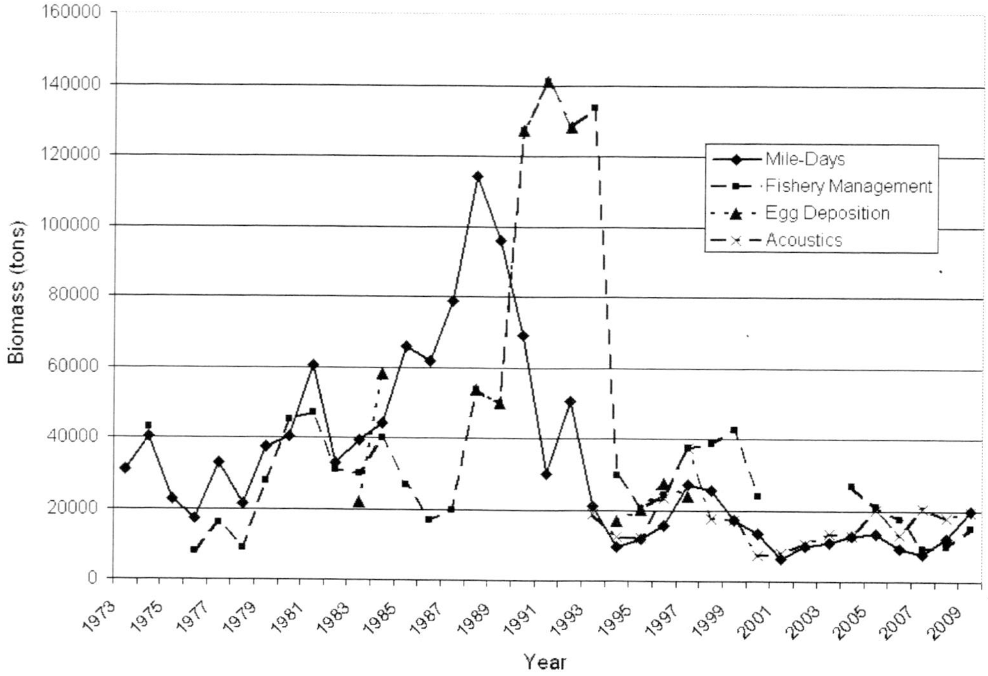

Figure 3. Comparison of four multi-decadal estimates of herring biomass in Prince William Sound. The Fishery Management and Egg Deposition values were identical in 1988-92.

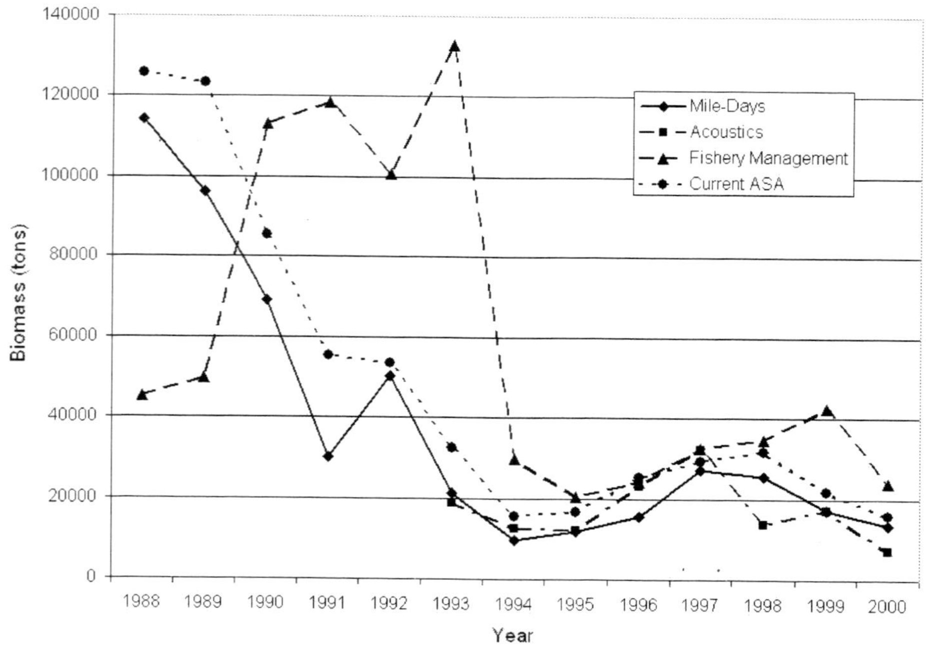

Figure 4. Comparison of various estimators from 1988 to 2000, all standardized to post fishery.

The ASA model has been revised several times since 1993 (Hulson et al. 2008). Back calculations using the ASA model in current use come closer to the historical biomass values of the hydroacoustics and the mile-days model, but still lag and overestimate slightly (Fig. 4). With the benefit of hindsight, the current ASA model follows the post-EVOS decline fairly well and is only a year late in the 1998 decline.

We believe the lag times we observed for detecting changes in stock biomass are a serious structural flaw in the preseason fishery forecast methods that are dominant in fisheries management today (Quinn and Deriso 1999). The possibility of a one year lag time when using preseason forecasting was identified briefly in the 1998 NRC review of fish stock assessment, but it was not pursued as an inherent flaw of the method. The possibility of longer lag times was not discussed. Certainly, lag times of five years were not considered possible. The collapse of a fish stock is likely when lag times of greater than one year are combined with large uncertainties in the biomass estimates.

Discussion

The most revealing aspect of our findings is the lag time errors. It is widely recognized that fisheries models have an inherent one-year lag-time, although the NRC did not focus on this source of error. In contrast, the hydroacoustic surveys provide a fishery-independent, real-time estimate of fishery abundance in the same season as the fishery so lag time is not an issue. The mile-days index provides a post-fishery measure of population size. However, the ASA forecasts are based on the previous year's data. Actual measures of subsequent mortality and recruitment are not included. This likely played a role in the original population collapse that started in 1989, since the oil spill added an undetected and immeasurable mortality to both juveniles and adults. The ASA models are especially sensitive to undetected higher mortality of adults, which explains how they can overestimate future levels of recruitment and stock biomass. However, the lag time is clearly more than one year, which warrants further investigation. It is likely that inherent averaging functions in the model structure add inertia that leads to greater lag.

Case History 2-An Indirect Estimate

Fishery-independent methods are typically considered to be those like hydroacoustics and research trawl surveys that directly assess stock abundance. However, ecosystem-based fishery management, as recommended by commissions such as the Pew Ocean Commission (2003) and the U.S Ocean Commission (2004), implies an understanding of ecosystem forcing functions, which we will show requires observation of complex and dynamic ecosystem processes. We illustrate this problem with the case history of the pink salmon (*Oncorhynchus gorbuscha*) run in Prince William Sound (PWS), Alaska. The difficulty of pink salmon forecasting has long been recognized (Adkinson, 2002; Pyper et al. 2002). Pink salmon spend only a little over one year at sea before returning as adults. There are two major problems that are associated with this short life cycle. One is the limited fishery-dependent data. For example, the robustness of age-structure analysis is based upon having a large adult biomass with several year-classes of fish as a reference. This is not available in the single age-class pink salmon fishery. Secondly, natural mortality becomes more critical and more variable with a short life cycle. Pink salmon returns are highly dependent on natural processes.

Historical information on the Prince William Sound pink salmon harvests extends back several decades (Fig. 5). A hatchery program was initiated in the late 1970's. The run has generally increased in magnitude since that time. It began an extreme even-odd year oscillation in 2000 that appears to have reversed phase in 2009. The last year, 2010, was a record harvest of 66.2 million fish. Currently, hatcheries release about one-half billion pink salmon fry into the waters of Prince William Sound between mid-April and mid-May of each year (Cooney 1993; Kline et al. 2008; Pyper et al. 2001; Thomas and Mathisen 1993). The survival of pink salmon fry has been described as food limited, with the large copepods of the genus *Neocalanus* as the primary prey (Cooney 1986, 1993). Research conducted by the Sound Ecosystem Assessment (SEA) program in the 1990s indicated that survival of the fry is dependent on the zooplankton food availability and predator abundance during the first 45-60 days of the spring plankton bloom, which is dominated by *Neocalanus* copepods (Cooney et. al 2001; Willette et al. 2001). Both survival and early growth rates of pink salmon were correlated with the duration of the *Neocalanus* spring bloom. The mortality of pink salmon fry during this time is a function of their growth rate, becoming less vulnerable as they increase in size and swimming ability so that they can avoid predators (D. Parker 1996; R. Parker 1968; Willette 2001). In addition to this accepted hypothesis for early, growth-related survival, Moss et al. (2005) argued that food conditions in the GOA were an important factor in pink salmon survival.

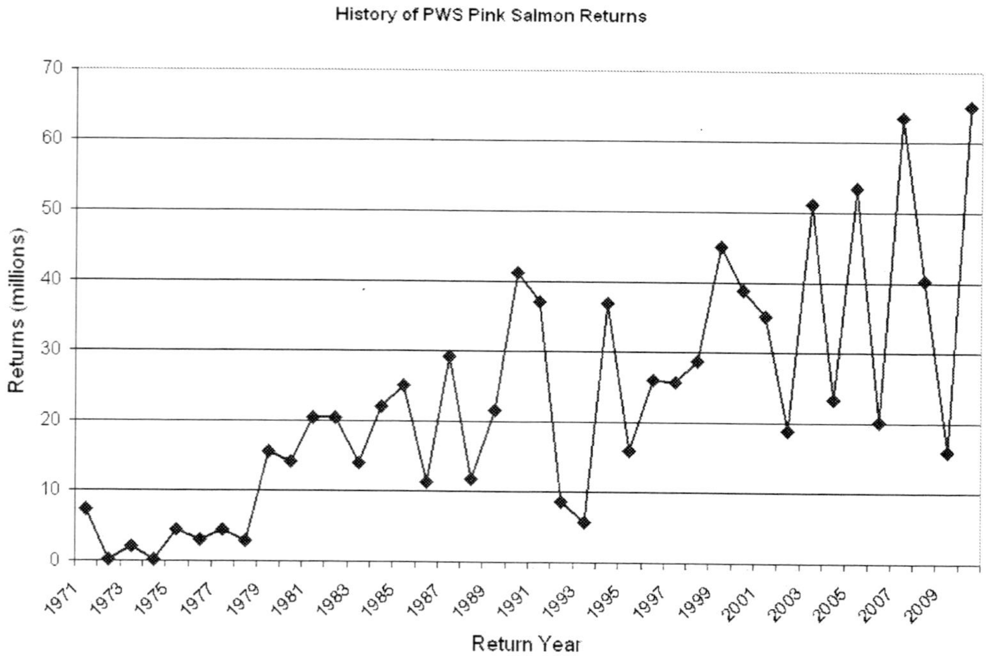

Figure 5. Historical pink salmon returns (harvests) in PWS showing increasing run size and associated annual variability, ending in an even/odd-year oscillation. Data available from Alaska Department of Fish and Game (e.g. www.cf.state.ak.us).

The most dominant pelagic fish predators in the PWS ecosystem are the walleye pollock (*Theragra chalcogramma*) and the Pacific herring (*Clupea pallasi*) (Thomas et al. 1997; Thorne 2008), which suggests that the marine assemblage may operate as a wasp waist trophic structure (Cury et al. 2003; Rice 1995). As one of the primary findings of SEA, Willette et al. (2001) showed that when *Neocalanus* abundance is low, pollock become piscivorous and by virtue of their high abundance become the dominant predator of pink salmon fry. Pacific herring, the second most abundant fish predator, also exhibits a similar prey switching behavior (Willette et al. 2001).

In response to the SEA findings, we initiated a program in 2000 to synoptically monitor the spring predator and prey densities of juvenile pink salmon with acoustics and nets. Our major goal was to improve upon the understanding of these growth-dependent survival and prey switching mechanisms, specifically to explain sources of variability in the survival of pink salmon fry. The surveys were conducted for seven years (2000-2006). The research and monitoring overlapped with the ongoing marking programs at the PWS hatcheries to determine pink salmon survival (continuous since the early 1990s), 15 years of our surveys of adult herring and pollock in PWS (Thorne 2008), and four years of surveys by the Northeast Pacific Global Ocean Ecosystems Dynamics (NEP-GLOBEC) program, which focused on pink salmon survival in the adjacent Gulf of Alaska during the summer months of 2001-2004 (Weingartner et al. 2002). A major outcome of our research is the demonstration of the multiple biological forcing functions that govern pink salmon survival and subsequent run size. This situation provides insurmountable obstacles to current fishery management approaches and calls for a new, truly ecosystem-based approach.

Methods

Our overall approach is described in Thorne (2008), which examined the role of walleye pollock in the Prince William Sound ecosystem. We made net tows for zooplankton during spring in six areas of Prince William for seven years. The number of samples varied annually from 20 to 63. The zooplankton sampling was conducted by making 50 m vertical tows using a 0.335-mm 0.5 m-ring net, following procedures of Cooney et al. (1995, 2001b). Our objective in the zooplankton sampling was to compare the average catch per tow of large copepods with subsequent pink salmon returns. Detailed pink salmon survival information is available since most PWS pink salmon are of hatchery origin and thermally marked (Hagen et al. 1995; Smoker and Linley 1997). There are four hatcheries that release pink salmon fry. The three northernmost of the hatcheries, Wally Noerenberg, Cannery Creek and Solomon Gulch, co-vary in marine survival (Pyper et al. 2001). The fourth, Armin F. Koernig (AFK), is located near the GOA and does not show the same survival trends as the others, but its returns are also the smallest of the hatcheries. The magnitude of the PWS pink salmon returns are driven primarily by the three northernmost hatcheries. Both total harvest and marine survival data are readily available on Alaska Department of Fisheries websites (e.g. www.cf.adfg.state.ak.us).

Our primary hypothesis was to test if the survival of pink salmon fry was the same during years of high and low *Neocalanus* abundance as measured by our catch per tow. We tested for annual differences in the log-transformed net catch per tow of large copepods as well as the significance of the regression of pink salmon survival on the average annual catch using standard "t" tables. In addition, since walleye pollock were identified as the major predator on juvenile pink salmon, we examined data on the overall abundance of walleye pollock. Sources included our own hydroacoustic surveys of walleye pollock in PWS (Thorne 2008) and data from the National Marine Fisheries surveys of the Gulf of Alaska (Guttormsen 2007).

Results

For the first five years of the study, the abundance of large copepods was significantly higher ($p \leq 0.05$) during even years (2000, 2002 and 2004) than during odd years (2001 and 2003). Adult pink salmon returns and marine survival initially varied in a similar manner as the large copepod abundance in the nursery year (Table 1). Pink salmon marine survival for even-year nursery conditions ranged from 5.3 to 10.4% and only 2.7 to 3.2% for the odd years, hence there was no overlap in the estimates. However, the correlation between survival and large copepod abundance was not significant over the five-year period ($r = 0.45$, $t = 1.18$, d.f. = 3, $p = 0.34$).

Table 1. Average net catch of large copepods in nursery year, associated pink salmon marine survival and returns (harvests)

Year	Average Large Copepod Catch per Tow	Subsequent Harvest (millions)	Average Marine Survival (%)
2000	1929	35.2	5.3
2001	584	18.9	2.7
2002	1403	51.1	10.4
2003	323	23.3	3.2
2004	841	53.3	8.8
2005	1128	20.2	4.6
2006	1081	55.0	7.1

Large copepod abundance in 2005 was significantly higher than the previous odd years and comparable to previous even year values. Large copepod abundance during the final year, 2006, was intermediate and not significantly different from any other year. However, pink salmon marine survival and returns continued the even/odd year oscillation even though large-copepod abundance in the nursery year no longer matched the pattern (Fig. 6).

There was a steady increase in pink salmon survival from the three northernmost hatcheries relative to food conditions over the first five years (Fig. 7). The average pink salmon survival for the three northernmost hatcheries divided by the average net catch of large copepods in the nursery year increased significantly over the 5-year period ($r = 0.98$; $p \leq 0.01$). The increasing survival trend between 2000 and 2004 is a primary reason for the lack of significant correlation between pink salmon survival and large copepod abundance ($r = 0.45$; Table 1). The combination of large copepod catch and year account for 96% of the variability (r^2) in pink salmon survival in nursery years 2000-2004.

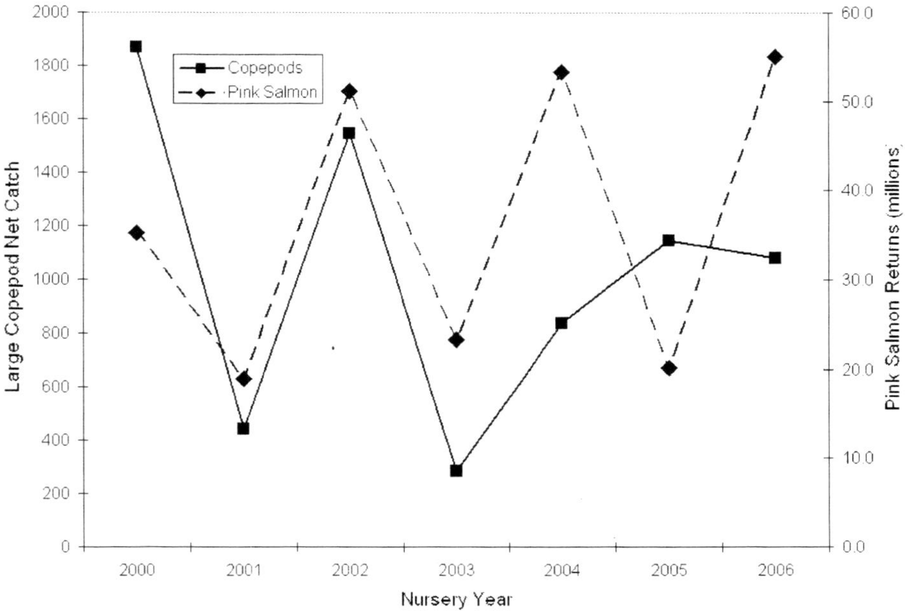

Figure 6. Comparison of the average large copepod net catch and subsequent returns of pink salmon for nursery years 2000 to 2006.

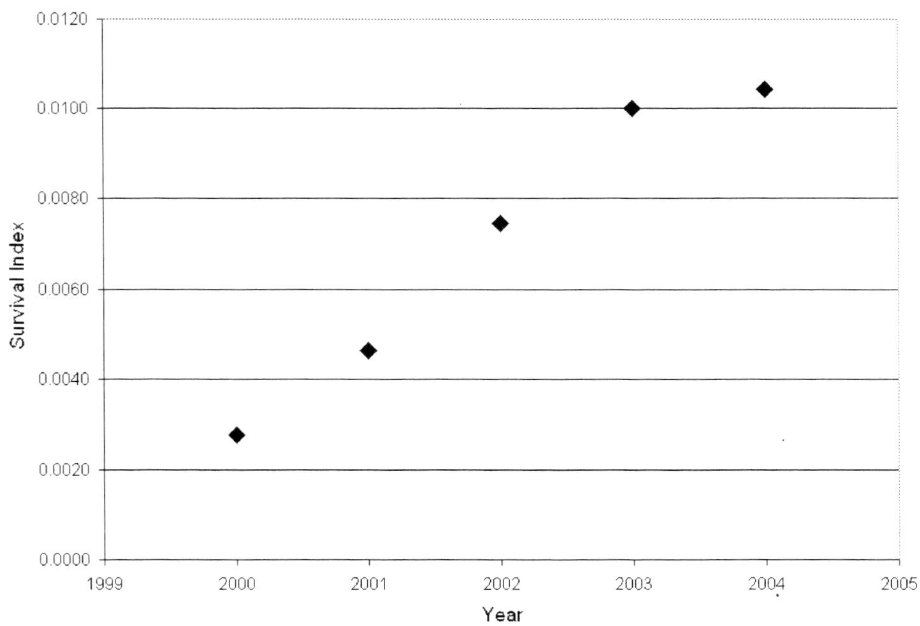

Figure 7. Increasing marine survival per large copepod in nursery year from 2000 to 2004. Index is average pink salmon survival from the 3 northernmost hatcheries divided by the average catch of large-bodied copepods in the nursery year.

Thorne (2008) reported a significant decline in the abundance of walleye pollock in Prince William Sound from 1998 to 2003. The decline was captured by hydroacoustic surveys conducted by the PWSSC in both PWS and in Port Bainbridge, which is just south of the pink fry outmigration corridor in the GOA (Fig. 8). It was also captured by the National Marine Fisheries Service hydroacoustic surveys of pollock in Shelikof Straits (Guttormsen 2007). It appears that the increase in the copepod-normalized pink salmon survival between 2000 and 2004 corresponds with a general decline in the abundance of walleye pollock, the major predator on juvenile pink salmon. The abundance of herring, the other major predator on juvenile pink salmon in PWS, has been at low levels for nearly two decades following the Exxon Valdez Oil Spill (Thorne and Thomas 2008).

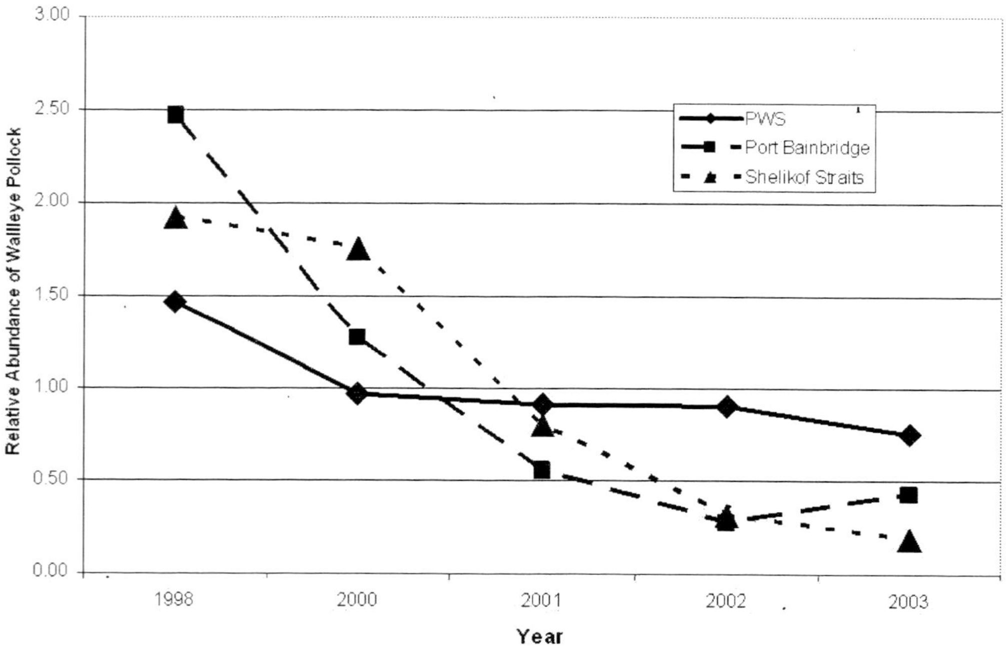

Figure 8. Walleye pollock abundance trends in Prince William Sound, Port Bainbridge and the Shelikof Straits in the Gulf of Alaska, 1998-2003. Annual estimates in each location relative to 1998-2003 average.

Discussion

The combination of large copepod catch and year account for 96% of the variability (r^2) in pink salmon survival in nursery years 2000-2004, and the increasing trend in survival during 2000-2004 appears to be the result of decreasing pollock abundance. However, no data explain the continuing two-year oscillation in the pink salmon survival after 2004. Marine survival and large copepod abundance varied in opposite directions the last two years of the study (Fig. 5). The dominant factor in pink salmon survival over the seven-year period is an even/odd year-oscillation, even though large copepod abundance was relatively high in 2005. The basis of the two-year cycle is puzzling. A two-year cycle is common in wild stocks, but unexpected in hatchery populations where annual releases are similar in magnitude. Hatchery runs in Japan and the Great Lakes have developed a two-year cycle (Radchenko et al. 2007). However, an examination of historic PWS runs shows an increasing trend of pink salmon abundance and annual variability, but no even-odd year cycle until recent years (Fig 5).

The two-year cycle in abundance of wild stocks of pink salmon has puzzled fisheries scientists for decades (Heard 1991, Radchenko et al. 2007). The obvious two-year signal that would be self-repeating for a two-year fish is cannibalism of the fry by the returning adults. Adult pink salmon do return along similar routes as their fry outmigration and at similar depth intervals, so their overlap could present a problem for fry survival if the adults fed on the fry. While cannibalism has been suggested by some researchers (see Ruggerone and Nielsen 2004), there have been no studies to quantify this behavior. The lack of evidence in PWS may be the result of the numbers involved, a fry biomass of around 250 mt in an adult salmon run biomass of 100,000 mt. Pink salmon survivals in recent years have exceeded 10%, meaning that each returning adult results from 10 or fewer fry. Consequently, if each returning adult eats 5 fry the impact on the average would be a 50% or greater reduction in the marine survival of the next year's return. Given a several week potential overlap, it would only take about 1 pink salmon fry consumed per week by each adult pink salmon to have the impact observed in the recent return data. Such a low concentration in stomach samples would be very difficult to detect.

Further support for cannibalism comes from the fact that the two-year cycle is most pronounced in the three northernmost hatcheries. The outmigration route to the GOA for the northernmost hatcheries exceeds 100 km, whereas the southern hatchery, AFK, is only 5 km from the Gulf. Unlike the other three hatcheries, AFK does not show the pronounced two-year cycle in pink salmon returns. It is possible that the alternating high and low *Neocalanus* abundance during the first four years of this study may have initiated the even-odd cycle, which was subsequently maintained by cannibalism.

Subsequent to our field study, the pattern of pink salmon returns to PWS deviated from the even/odd year cycle that we documented (Fig. 5). The 2008 pink salmon return to PWS was 40.2 million fish, much higher than previous even years, and dominated by the three northernmost hatcheries (www.cf.adfg.state.ak.us). While there has been no comprehensive monitoring of likely forcing mechanisms, the high returns in 2008 suggest that cannibalism was not a major factor. Water temperatures during the 2007 nursery year were unusually cold and may affected one or more of the prey and predators of the pink salmon juveniles and forced pink salmon fry into a nearshore distribution, thus minimizing encounters with returning adults. The 2009 return was unexpectedly low at 16.5 million, perhaps the result of cannibalism from the large 2008 return, while the 2010 return was a record high most likely the result of limited cannibalism from the low 2009 return. The completely unpredicted returns from these past three years underscore the complexity and the need for major improvements in monitoring of biological forcing in order to predict marine survival.

General Discussion and Conclusion

These two case histories illustrate two different examples of fishery-independent data. One is a direct and precise estimate of the biomass of adult herring, and the other an indirect measure of the conditions in the nursery area, which could influence survival of the fry to adults. The first case history argues that repeated surveys over time on spawning aggregations is the only precise way to estimate adult herring biomass and show change over time, and that repeated ocean surveys are the only way to be sure that the measurements are representative of the stock. The stochastic data on herring biomass were not only consistent with large anthropogenic events, such as commercial fishing and an oil spill, but they show clear associations with predator assemblages in the ecosystem, Steller sea lion, humpback whales and several seabirds (Thorne and Thomas 2008). This is attributed to the elimination of lag times and inaccuracies that prevent the previous deterministic sources of data from being representative of the stock size. Perhaps some of the blame that has been placed on the single species approach could be better directed at the quality of the fisheries independent data being used in management and research. We conclude that a return to the rigor of collecting quality empirical data on adult fish abundance is a prerequisite to improving our understanding of the marine trophic structure and ecosystem.

The second case history shows that the survival of juvenile pink salmon is a function of several biological forcing factors in the spring bloom, some that are seasonally variable in abundance, such as the *Neocalanus* copepods and pteropods. Just the fact that any one or combination of these four factors can dominate the survival of pink salmon in any particular year results in 35 possible combinations of biological forcing, which is clearly beyond the purview of current prediction, and certainly is not the focus of current monitoring efforts. Although the study of physical forcing factors, such as water temperature, wind and currents may explain stratification, which affects the prey and predators of pink salmon, such physical forcing is obviously too indirect to serve as an adequate surrogate for the forecasting of adult pink salmon survival when the dominant forcing is the product of at least 35 possible combinations in abundance of predators and prey. The bottom line is that measurement of the variables in such a complex mechanistic model of survival does require considerably more directed observational effort than government sponsorship has supported in the past. If we are to advance the science and improve predictions of trophic structure function and specific stock dynamics, a new understanding of such problems will have to emerge from fisheries managers and sponsors of research.

REFERENCES

Adkison, M.D. (2002) Preseason forecasts of pink salmon harvest using Bayesian model averaging, *Alaska Fish Res. Bull.* 9:108.

Anon. 1998. Improving Stock Assessment. National Research Council.National Academy Press. Washington D.C. 177 pages.

Becker, K.E. and E.D. Biggs 1992. Prince William Sound herring spawn deposition survey manual. Regional Information Report 2A92-05, 2C92-02, Alaska Dept. of Fish and Game, Anchorage, AK 35p.

Brown, E.D. 2007. Pacific Herring, pp 290-299 in *Long-term Ecological Change in the Northern Gulf of Alaska*, Robert Spies (ed), Elsevier, Amsterdam, 589 p.

Casey, J. M., and R. A. Myers. 1998. Near extinction of a large widely distributed fish. *Science* 281:690-692.

Cooney, R.T. (1986) Zooplankton, In: *The Gulf of Alaska, physical environmental and biological resources*, D.W. Hood and S.T. Zimmerman (eds) U.S Depart. of Commerce, Minerals Management Service, Anchorage, pp. 285-304.

Cooney, R.T. (1993) A theoretical evaluation of the carrying capacity of Prince William Sound, Alaska, for juvenile Pacific salmon, *Fish. Res*. 18:77-87.

Cooney, R.T., Willette, T.M., Sharr, S., Sharp, D. and Olsen J. 1995. The effect of climate on North Pacific pink salmon (*Onchorynchus gorbuscha*) production: examining some details of a natural experiment, In: *Climate Change and Northern Fish Populations*, R.J. Beamish (ed.), *Can Spec. Publ. Fish Aquat*. Sci. 121, pp. 475-482.

Cooney, R.T., Allen, J.R., Bishop, M.A., Eslinger, D.L., Kline, T., Norcross, B.L., Mcroy, C.P., Milton, J., Olsen, J., Patrick, V., Paul, A.J., Salmon, D., Scheel, D., Thomas, G.L., Vaughan, S.L. and Willette T. M. 2001. Ecosystems controls of juvenile pink salmon (*Onchorynchus gorbuscha*) and Pacific herring (*Clupea pallasi*) populations in Prince William Sound, Alaska, *Fish. Oceanogr*. 10 (Suppl 1):1-13.

Cury, P., L. Shannon and Y. Shin (2003) The functioning of marine ecosystems, In: *Responsible Fisheries in the Marine Ecosystem*, M. Sinclair and G. Valdimarson, (eds) Cambridge, MS: CABI Publishing. p. 103-124.

Gunderson, D. 1993. *Surveys of Fisheries Resources*. John Wiley & Sons. New York. 347 pages

Guttormsen, M.A. (2007) Results of the February-March 2007 echo integration-trawl surveys of walleye pollock (*Theragra chalcogramma*) conducted in the Gulf of Alaska, Cruises MF2007-1 and MF2007-4, AFSC. Processed Rep. 2007-07, 75p. Alaska Fish. Sci. Cent., NOAA, Natl. Mar. Fish. Serv., 760 Sand Point Way N.E. Seattle 98115.

Hagen, P., Munk, K., Van Alen, B. and White, B. (1995) Thermal mark technology for inseason fisheries management: a case study, *Alaska Fish. Res. Bull.* 2:143-155.

Heard, W.R. (1991) Life history of pink salmon (*Oncorhynchus gorbuscha*), In: *Pacific Salmon Life Histories*, C. Groot and L. Margolis (eds), UBC Press, Vancouver, B.C. Canada, pp. 119-230.

Hilborn, R. 2006. Faith-based Fisheries. Essay: Professional Issues. *Fisheries* 31(11): 554-555.

Hulson, P-J. F., S.E. Miller, T.J. Quinn II, G.D. Marty, S.D. Moffitt, and F. Funk 2008. Data conflicts in fishery models: incorporating hydroacoustic data into the Prince William Sound Pacific herring assessment model. *ICES Journal of Marine Science*, 65: 25–43.

Kline, T.C., Boldt, J.L., Farley, E.V. Jr., Haldorson, L.J. and Helle, J.H. (2008) Pink salmon (Oncorhynchus gorbuscha) marine survival rates reflect early marine carbon source dependency, *Prog. Oceanogr*. 77:194-202.

Meyers, R. A. and B. Worm. 2003. Rapid world-wide depletion of predatory fish communities. *Nature* 423:280-283.

Moss, J. H., Beauchamp, D.A., Cross, A.D., Myers, K.W., Farley E.V. Jr., Murphy, J.M. and Helle, J.H. (2005) Higher marine survival associated with faster growth for pink salmon (*Oncorhynchus gorbuscha*), *Trans. Am. Fish. Soc.* 134: 1313-1322.

Parker, D. (1996) The feeding ecology and growth of juvenile pink salmon (*Oncorhynchus gorbuscha*) in Prince William Sound, Alaska, In: *Proceedings of the 17th Northeast Pacific Pink and Chum Salmon Workshop*, 1-3 March 1995, Bellingham, Washington, Northwest Indian Fisheries Commission, Washington Department of Fish and Wildlife, Olympia, pp. 23-39.

Parker, R.R. (1968) Marine mortality schedules of pink salmon of the Bella Willette, T.M. (2001) Foraging behavior of juvenile pink salmon (*Oncorhynchus gorbuscha*) and size dependent predation risk, *Fish. Oceanogr.* 10 (Supplement 1):110-131.

Pauly, D . 2009. Beyond duplicity and ignorance in global fisheries* *Scienta Marina.* 73(20):215-224.

Petersen, C.G.J. 1894. On the biology of our flatfishes and on the decrease of our flatfish fisheries. App. IV. On the labeling of living plaice. *Appendix Danish Biol. Stat.* 4:140-143.

Peterson, C.H., Rice, S.D., Short, J.W., Esler, D., Bodkin, J.L., Belliachey B.E. and Irons, D.B. 2003. Long-term ecosystem response to the Exxon Valdez Oil Spill. *Science* 302:2082-2086.

Pew Ocean Commission 2004. *Americas Living Oceans-Charting a Course for Sea Change*, PEW Oceans Commission, May 2003

Pyper, B.J., Mueter, F.J., Peterman, R.M., Blackbourn, D.J. and Wood, C.C. (2001) Spatial covariation in survival rates of Northeast Pacific pink salmon (*Oncorhynchus gorbuscha*), *Can. J. Fish. Aquat.* Sci. 58:1501-1515.

Quinn, T.J. II and Deriso, R.B. 1999. *Quantitative fish dynamics*, Oxford University Press, NY 542 p

Quinn, T.J. II, G.D. Marty, J. Wilcock and M. Willette 2001. Disease and population assessment of Pacific herring in Prince William Sound, Alaska. In F. Funk, J. Blackburn, D. Hay, A.J. Paul, R Stephanson, R. Toresen and D. Witherell (eds) *Herring Expectations for a new Mellennium*, pp. 363-379, Alaska Sea Grant College Program, AK-SG-01-04.

Quinn, T. J. II. 2003. Ruminations on the development and future of population dynamics models in fisheries. Natural Resource Modeling 10(4): 341-346.

Radchenko, V.I., Temnykh, O.S. and Lapko, V.V. (2007) Trends in abundance and biological characteristics of pink salmon (*Oncorhynchus gorbuscha*) in the North Pacific Ocean, *N. Pac. Anadr. Fish. Comm. Bull.* 4:7-21.

Rice, J. (1995) Food web theory, marine food webs and what climate change may do to northern marine fish populations, In: *Climate Change and Northern Fish Populations*, R.J. Beamish (ed.) *Can. Spec. Publ. Fish. Aquat. Sci.* vol. 121, pp. 561-568.

Ruggerone, G.T. and Nielsen, J.L. (2004) Evidence for competitive dominance of pink salmon (*Oncorhynchus gorbuscha*) over other salmonids in the North Pacific Ocean, *Rev. Fish Biol. Fish.* 14:371-390.

Schweigert, J., C.W. Haegele and M. Stocker 1990. Evaluation of Sampling Strategies for Scuba Surveys to Assess Spawn Deposition by Pacific Herring *North American Journal of Fisheries Management* 10: 185-195

Simmonds, J. and D. MacLennon 2005. *Fisheries Acoustics: theory and practice, 2nd Ed.*. Blackwell Science, 429 p

Smoker, W. W. and Linley, T.J. (1997) Are Prince William Sound salmon hatcheres a fool's bargain, (*Alaska Fish. Bull.* 4:74-78.

Thomas, G.L. and Mathisen, O. (1993) Special Issue: Biological interactions between enhanced and wild salmon in Alaska, *Fish. Res.* 18(1-2): 1-159.

Thomas, G.L. and R.E. Thorne 2001. Night-time Predation by Steller Sea Lions. *Nature* 411:1013.

Thomas, G. L. and Thorne, R.E. 2003. Acoustical-optical assessment of Pacific herring and their predator assemblage in Prince William Sound, Alaska. *Aquatic Living Resources.* 16:247-253.

Thorne, R.E. and Thomas, G.L. 2008. Herring and the "Exxon Valdez" oil spill: an investigation into historical data conflicts. *ICES Journal of Marine Science* 65:44-50.

Thorne, R.E. (2008) Walleye pollock as predator and prey in the Prince William Sound ecosystem. In: G.H. Kruse, K. Drinkwater, J.N. Ianelli, J.S. Link, D.L. Stram, V. Wespestad and D. Woodby (eds), *Resiliency of gadid stocks to fishing and climate change*, Alaska Sea Grant, University of Alaska, Fairbanks, pp. 289-304.

U.S Ocean Commission 2004. *An Ocean Blueprint for the 21st Century*, US Commission on Ocean Policy.

Walters, C. 2003. Folly and fantasy in the analysis of spatial catch real data. *Can. Journal. Fish. Aquatic Sciences* 60:1433-1436.

Weingartner, T.J., Coyle, K., Finney, B., Hopcroft, R., Whitledge, T., Brodeur R., Dagg, R., Farley, E., Haidvogel, D., Haldorson, L., Hermann, A., Hinckley, S., Napp, J., Stabeno, P., Kline, T., Lee, C., Lessard, E., Royer, T. and Strom S. (2002) The Northeast Pacific GLOBEC Program: Coastal Gulf of Alaska, *Oceanography* 15:48-63.

Willette, T.M., Cooney, R.T., Patrick, V., Mason, D. M., Thomas G.L. and Scheel, D. 2001. Ecological processes influencing mortality of juvenile pink salmon (*Onchorynchus gorbuscha*) in Prince William Sound, Alaska, *Fish. Oceanogr.* 10 (Suppl 1):14-41.

Worm, B., Hilborn, R., Baum, J.K., Branch, T.A., Collie, J.S., Castello, C., Fogarty, M., Fulton, E.A., Huthchings, J.A., Jennings, S., Jensen, O.P., Lotze, H.K., Mace, P.M., McClanahan, T.R., Minto, C., Palumbi, S.R. Parma, A.M., Ricard, D., Rosenberg, A.A., Watson, R. and Zeller D. Rebuilding Global Fisheries. *Science* 325 July 2009. 578-585.

In: Fishery Management
Editor: Janice S. Intilli

ISBN: 978-1-61209-682-7

Chapter 9

FISHERIES MANAGEMENT: EVALUATING THE RECOVERY PLAN FOR EUROPEAN HAKE FISHERY (SOUTHERN STOCK)

Dolores Garza-Gil, Gonzalo Caballero-Miguez, Manuel M. Varela-Lafuente and Juan Surís-Regueiro
Departament of Applied Economics; University of Vigo; Lagoas-Marcosende s/n; 36310 Vigo; Spain

ABSTRACT

Since the European hake stock is overexploited and presents reduced reproductive capacity, the European Commission (EC) drew up a Recovery Plan, which aims to return this species within biological safety limits in ten years (EC, 2005). In this chapter, we will evaluate this management plan, put forward different scenarios and analyse how levels of biomass and catches equilibrium would be affected. A model for considering the size-selectivity of the fishing gears in the dynamic of resource is proposed. The results show that if the less conservationist fleet improves its degree of fishing selectivity, the level for the marine resource will increase and it would possible to place the stock within the limits of safety in ten years (aim of the recovery plan for this stock) and to reach the optimum values. If not, a greater reduction in the TACs should be applied for reaching the limits of biological safety.

1. INTRODUCTION

In the analysis of fisheries in which various fleets using different gear carry out their activity it is common to assume that such activity influences the net growth of the marine resources through the catches, while the natural growth function depends on the biomass and environmental conditions, generally stable and constant (e.g., Gallastegui, 1983; Armstrong, Clark, 1997; Garza-Gil, 1998; Costa-Duarte et al, 2000; Bjorndal, Lindroos, 2003). However, some fishing gear can alter the composition of fish populations or modify the recruitment rate (Lleonart, Recasens, 1996) and, therefore, it is advisable to include the degree of their selectivity in the analysis. Fishing selectivity can cause two different problems: by-catch (inter-species selectivity) and catches of 0 age or small individuals of the objective species (intra-species selectivity or size-selectivity). The former has been dealt with extensively in economic literature (e.g., Boyce, 1996; Prellezo, Gallastegui, 2003; Turner, 1997; Ward, 1994). However, insofar as we are aware, the problem of size- selectivity has only been partially studied by Escapa and Prellezo (2003), who analyse the conditions of stationary equilibrium in a fishery in which fishing technology influences the natural growth function (assuming it is of the quadratic type).

Given that one of the fleets operating in the fishery we are studying (European hake) captures mainly individuals of a lower size than that corresponding to sexual maturity, we are going to focus on the latter type of selectivity. Unlike Escapa and Prellezo, we are assuming a natural growth function of the Cushing type (given that it is the one that presented best econometric results) and will pose a parameter for catching the each technology's selectivity. The particular issue with which this paper is concerned is how the degree of size-selectivity may affect the resource.

Since the hake stock is overexploited and presents reduced reproductive capacity, the European Commission (EC) drew up a Recovery Plan approved by the Council in October 2005 and which came into effect in 2006, which aims to return this species within biological safety limits in ten years (EC, 2005). The recovery plan includes the establishment of decreasing Total Allowable Catches (TACs), the reduction of total effort in the fishery and the need for that one of the fleets operating in this fishery (trawling fleet) improve its degree of selectivity by concentrating on individuals of a larger size. We will use the model for evaluating the recovery plan, put forward different scenarios and analyse how levels of biomass equilibrium would be affected.

The study is organised as follows: in section 2 we present the model; in section 3 we comment the data from the fishery; while in section 4 the equilibrium results are shown and the recovery plan is evaluated; lastly, the most significant conclusions will be set out in section 5.

2. THE MODEL

In order to provide a model for the effect of the fishing activity of each fleet on the fish population and following Garza and Varela (2007), we define a parameter γ_i ($0 \leq \gamma_i < 1,\ i = 1,2$) which establishes each technology's level of fishing selectivity (through catch composition) in the following way:

$$\gamma_i = \frac{N_{Ji}}{N_i} \qquad \text{with } 0 \leq \gamma_i < 1; \ \ i = 1,2 \qquad [1]$$

where N_{Ji} represents the number of juvenile individuals (individuals which have not yet reached sexual maturity size) caught by the fleet *i*; and N_i the total number of individuals caught by said fleet.

If the fleet technology *i* is very selective catching mature individuals that spawned some time, γ_i tend to the null value. This fleet will have less negative effects on the marine resource, and will be more conservationist with the resource. On the contrary, if a fleet catches a relatively high number of juveniles (which will not reach the spawning age), the value of the parameter γ_i will tend to one. This technology will affect the natural growth function negatively, and then this scarcely selective fleet will be less conservationist with the resource.

The intensity of the effect fishing technology has on the growth of the resource also depends on the catches made using each method in relation with global catches in the fishery. For it, we define a θ variable which includes the selectivity parameter of each fleet and the catches relating to each one of them in respect of the total catches generated in the fishery in the following way:

$$\theta_t = 1 - \sum_i \gamma_i \frac{h_i(X_t, E_{it})}{h(X_t, E_t)} \qquad \text{with } 0 < \theta \leq 1;\; i = 1,2 \qquad [2]$$

where h_i denotes the fleet catches i, h the total catches in the fishery, E_i is the effort exercised by the fleet i and X the fish stock. θ will approximate one the lower the global effect of both technologies on the marine resource. And the natural growth function of the marine resource (F) will be modified by θ and is represented as:

$$G(X_t, \theta_t) = \theta_t \, \mathrm{F}(X_t) \qquad [3]$$

Under these assumptions, and in a context of centralised resource management (as could be the case of the European Union), the problem of optimisation which it must resolve consists of choosing the effort levels (E_i, $i=1,2$) which maximise the trend of the profits that are generated in the fishery discounted at the initial moment, and including the fish population dynamics in the problem, that is,

$$\underset{E_i}{Max} \sum_i \int_0^{\infty} \{p_i h_i(E_i(t), X(t)) - w_i E_i(t)\} e^{-\delta t} \partial t \qquad i = 1,2$$

$$s.a. \quad \dot{X} = \frac{\partial X}{\partial t} = G(X, \theta) - \sum_i h_i(E_i(t), X(t))$$

$$0 \leq E_i(t) \leq E_{i\,\max}$$

$$0 < x(t)$$

where p and w denote, respectively, the unitary price of the catches and the cost per unit effort exercised by each one of the fleets; and δ represents the social discount rate. $\dot{X}$ reflects fish population dynamics, that is, the net marine resource growth, once the catches have been discounted.

The associated equilibrium problem is determined from the following equations (Garza and Varela, 2007):

$$(p_i - \mu) \frac{\partial h_i(.)}{\partial E_i} + \mu \frac{\partial G(.)}{\partial E_i} = w_i \qquad [4]$$

$$\dot{\mu} - \delta\mu = -\left[(p_i - \mu)\sum_i \frac{\partial h_i(.)}{\partial X} + \mu\frac{\partial G(.)}{\partial X}\right] \qquad [5]$$

$$\dot{X} = G(X) - \sum_i h_i(.) \qquad [6]$$

where μ represents the shadow price or social value of the natural resource. From the system of equations [4]-[6] the stationary equilibrium solutions for the fish biomass, effort and social value of the marine resource variables are obtained. These values will make it possible to maintain the sustainable operation of the fishery.

3. The Southern Hake Fishery

The fishery under study is the southern hake stock (*merluccius merluccius*) in VIIIc and IXa ICES zones. The species is caught using two kinds of gear: trawl and artisanal gears. These fishing methods affect the hake population differently. Trawling, although it catches all age segments, acts intensely on young individuals (smaller than 30cm), preventing them from reaching adulthood. Artisanal fleet targets mainly mature individuals of between 30 and 80cm, causing less negative effect on the hake population[1]. In Table 1 we can see the average catch composition of both fleets for the period 2001-2005.

Table 1. Average number of individuals (in thousands) per size (cm) caught by the trawl and artisanal fleets. 2001-05.

Size groups	Trawl	Artisanal
5-15	1,521	0
16-30	5,416	342
31-45	2,633	1,104
46-60	1,902	5,571
61-75	24	4,678
+76	0	247
Total individuals	15,859	11,942

Source: ICES, Working Group on the Assessment of Southern Demersal Stocks.

It can be observed that the percentage of adult individuals in the trawl catches is lower than 40%, while in the case of the artisanal catches this reaches 90%. Trawl technology, therefore, insofar as its present technical characteristics are concerned, is preventing a significant number of individuals from reaching sexual maturity in adulthood. Consequently, this fleet is altering the composition of the fish population and limiting their reproduction capacity, thus affecting the generation of future profits in the fishery. This means that in this fishery not only is it important to depict the effect of global catches on resource dynamics, but

[1] The size at which the species is considered to have reached sexual maturity stands at 31-35cm for males and 40-45cm for females. The minimum catch size is 27cm. More biological information on hake stock can be found in Trujillo et al (1991).

also the different effect that both technologies are having on the marine resource, given the ways in which they act on it. Artisanal fishing, on the other hand, and on account of its specific characteristics, is a relatively more selective method, especially in a fishery such as the one we are analysing, in which the hake individuals are distributed at different depths according to their age and size (the deeper the water, the older and bigger the individuals).

Table 2. Ibero-Atlantic hake: stock, catches and effort. Period 1985-05. (Data in thousands)

Year	X (tons)	h_1 (tons)	h_2 (tons)	E_1 (fishing days)	E_2 (fishing days)
1985	54.4	6.9	10.7	64.4	71.3
1986	51.7	8.8	14.2	56.6	72.4
1987	43.4	7.7	14.4	35.6	43.7
1988	31.7	8.2	10.8	42.7	43.1
1989	27.6	7.6	9.6	35.2	64.6
1990	26.7	5.8	10.4	29.4	76.4
1991	25.6	6.4	9.9	40.0	73.0
1992	23.6	6.6	7.1	46.4	99.5
1993	22.7	6.4	6.7	35.9	82.2
1994	22.8	6.0	6.7	34.4	50.3
1995	22.9	6.1	7.7	34.9	46.2
1996	20.8	4.1	7.4	27.7	71.4
1997	17.6	3.8	6.0	26.9	88.3
1998	15.3	6.7	5.5	31.9	105.4
1999	14.0	5.5	4.4	25.0	71.1
2000	12.1	4.9	3.7	25.9	36.0
2001	11.8	4.3	3.4	22.7	33.1
2002	12.6	4.4	3.3	21.6	32.1
2003	14.4	4.0	3.3	19.4	21.6
2004	14.9	5.4	2.2	19.5	21.4
2005	15.0	5.8	2.1	--	--
2006	15.3	7.4	2.6	--	--

Sub-indexes 1 and 2 represent, respectively, the trawl and artisanal fleets; h_i catches from i, E_i fishing effort exercised by i

Source: Own compilation from ICES and personal interviews.

The European Union, within the framework of the Common Fisheries Policy, has regulated the fishery through TACs and minimum catch sizes since 1987. Furthermore, a closed list of vessels from each fleet defined on the basis of historical rights regarding participation in the fishery exists.

Regarding the economic parameters, the price per unit output for each fleet, we have the annual average data (in constant units) which range from around 4,565 to 7,482 euros per tonne of hake landed by the trawl and artisanal fleets, respectively. The costs per unit effort (in constant units) are estimated at 373 and 419 euros per trawl and artisanal fishing day, respectively[2]. Finally, we will use the average yield of long-term government bonds as the social discount rate (discounting the rate of inflation) for the period 1985-2005, situated at an annual rate of around 5%.

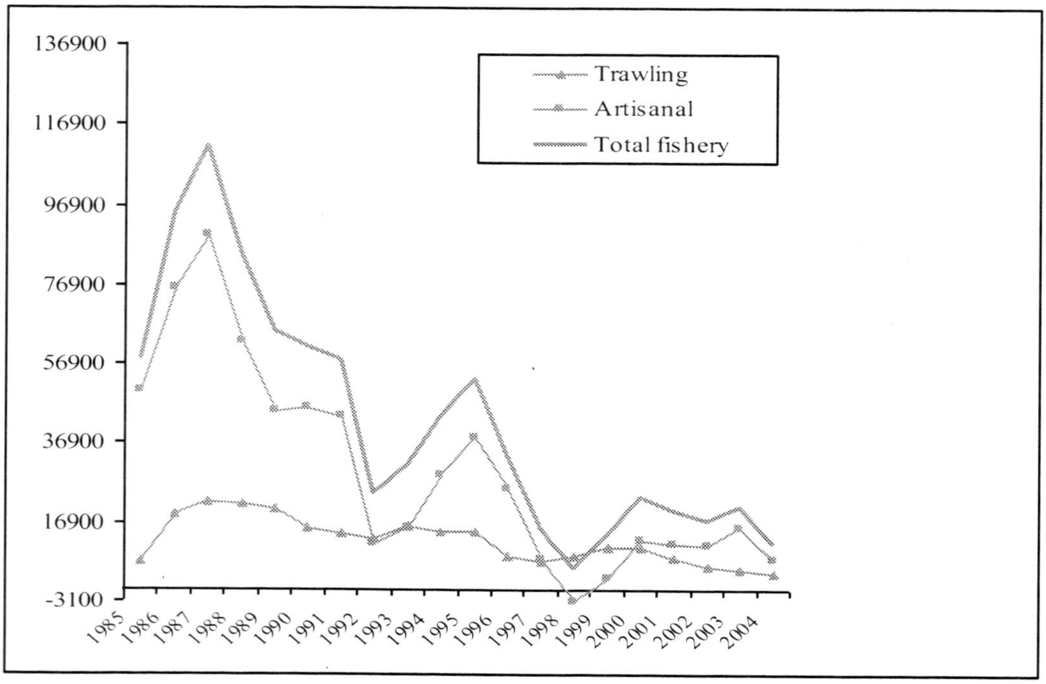

Source: Own compilation from Economic Assessment of European Fisheries (several years).

Figure 1. Profits (1 000 Euros) of trawling and artisanal fleets. 1985-2005.

4. Estimation Results

In the empirical section, on one hand, we will estimate the equilibrium values for the study case and, on the other hand, we will evaluate the recovery plan approved by the European Commission estimating the number of years necessary for reaching the biological safety level.

4.1. Equilibrium Values

In order to obtain the equilibrium values, it is necessary to estimate previously the respective parameters for the natural growth function and for the fishing technology of both fleets, as well as the selectivity parameter for each of them. With relation to the natural

[2] Average prices and costs correspond to the period 1995-05 in constant units for 2006 and have been estimated on the basis of the Economic Assessment of EU Fisheries (1998-2006).

growth of the European hake stock (southern stock), the function provides the best econometric results is the following[3]:

$$F(X) = X_{t+1} + h_t = aX_t^b \qquad [7]$$

With $a > 0$ and $b < 1$.

The results of the econometric estimations are shown in Table 3. Therefore, the growth function will be:

$$F(X) = 8.9263\ X^{0.82310} \qquad [8],$$

and the modified function will be represented by the following expression:

$$G(X,\theta) = \theta\ F(X) = \tilde{a}X^{0.82310} = 8.9263\ \theta\ X^{0.82310} \quad \text{with } \tilde{a} < a \qquad [9],$$

where $\tilde{a} = \theta\ a$ is the biological parameter modified by the effect of the activity of the different fleets.

Table 3. Natural growth function of the southern hake stock and fishing technology. Econometric estimations.

Natural fuction	Values	Fishing technology	Trawling values	Artisanal values
Ln a	2.1890* (7.519)	*Ln A*	3.0363* (2.463)	
b	0.82310* (28.340)	α_1	0.37744* (1.692)	0.12609** (1.238)
R^2	0.9780	β_1	0.17164* (1.077)	0.74163* (6.673)
F-statistic	802.931	R^2	0.5818	0.7920
Harvey test	0.016	F-statistic	14.214	
Q(1)	1.61	Harvey test	1.116	
		LM-statistic		8.3678
		Q(1)	0.76	6.28

t-ratios between brackets. * Indicates a significance level with 95% probability. *Ln* denotes Neperian logarithm. The Harvey test is the heterocedasticity test used when the constant is significant; LM (Lagrange multiplier) is the statistic of the Jarque-Bera heterocedasticity test used when the constant is not significant; Q (Box-Pierce) the one used in the self-correlation test.

[3] The quadratic (frequently used in economic fishery literature) and logistic forms were estimated. The first case presented a high degree of multicollinearity between the explicative variables. In the latter case, there was a lower level of statistical significativeness in the value of parameters than that obtained in [8]. On other hand, a growth function of the Beverton-Holt type could be applied to the selectivity problem. However, we have been unable to obtain biological information on the size and weight of the individuals of each group for a sufficiently long time series (we only obtained the data showed in Table 1).

The fishing technology commonly used in fishery models is of the Cobb-Douglas type[4]. In the hake fishery, the functional form that generated the best econometric results, after taking the Neperian logarithm, is[5]:

$$Ln\, h_{it} = Ln\, A + \alpha_i\, Ln\, E_{it} + \beta_i\, Ln\, X_t \qquad i = 1,2 \qquad [10]$$

The results of the estimations of expression [10] for the trawling and longline fleets are shown in Table 3. Therefore, the fishing technology of each fleet will be represented, respectively, by the following expressions:

$$h_{1t} = 20.82804\, E_{1t}^{0.37744}\, X_t^{0.17164} \qquad [11]$$

$$h_{2t} = E_{2t}^{0.12609}\, X_t^{0.74163} \qquad [12]$$

By substituting the information in Table 1 in expression [1] it is possible to estimate the value of the selectivity parameter for each fleet. In the case of the hake population, the age of sexual maturity is not reached up until a size of 30-45 cm (male and female, respectively); therefore, the group of juveniles will be made up of the individuals which do not reach such size. On this basis, the value of the selectivity parameter corresponding to trawl and artisanal activity is, respectively, as follows:

$$\gamma_1 = 0.60343 \qquad [13]$$

$$\gamma_2 = 0.12109 \qquad [14]$$

Substituting the parameter values in the equation system [4]–[6], the solutions for the stock levels are obtained, as are their shadow price and the effort exercised by each fleet. On the other hand, insofar as the trawl fleet could improve its level of selectivity, as the ICES and the EC propose in the management recommendations for this fishery (by increasing, for example, net and/or codend mesh sizes)[6] and so reduce the negative effects of its activity on the hake biomass, it would be interesting to pose different γ_1 values. In Table 4, the results obtained for different γ_1 values are shown.

As can be seen, as the selectivity of the trawl fleet improves $(\gamma_1 \rightarrow 0)$, the value of the modified biological parameter increases $(\tilde{a} = \theta a)$ and the optimum hake biomass level and its shadow price increases. Furthermore, the level of effort of the trawl fleet increases, its catches increasing, while the effort of the artisanal fleet decreases (as does the global effort in the fishery) as the selectivity of the less conservationist fleet improves and, therefore, the negative effect of the activity of said fleet on the natural growth of the resource is reduced.

---

[4] We would underline, among others, the works by Armstrong & Clark (1997), Bjorndal & Lindroos (2003), Clark (1990), Costa-Duarte et al (2000), Garza-Gil (1998), Gallastegui (1983), Surís (1993) and Weitzman (2002).

[5] In fishery economics, the constant of the Cobb-Douglas function is called the catchability coefficient (A = q > 0) and includes the catches made by each fleet per unit of stock and fishing effort.

[6] It is technologically possible and would not imply an additional cost, relevant bearing in mind that the repair and change of nets is a common cost for the fishery sector.

Table 4. Stationary solutions for different γ_1 and $\gamma_2 = 0.1$ values

γ_1	X (tons)	$\tilde{a}$	E1 (fishing days)	E2 (fishing days)	h1 (tons)	h2 (tons)	μ (euros)
0.6	36,143	6.91027	3,138	18,906	2,770	8,310	2,431
0.5	37,116	7.11883	4,103	17,592	3,079	8,339	2,548
0.4	37,262	7.33457	5,729	16,343	3,495	8,345	2,571
0.3	42,499	7.59560	7,321	12,682	3,924	8,911	3,499
0.2	43.014	7.63321	7,743	12,041	4,017	8,932	3,507

1: trawl; 2: artisanal. With p_i=4.57 thousand euros/tons; p_j=7.48 thousand euros/tons; w_i=0.37 thousand euros/day; w_j=0.42 thousand euros/day; δ=0.05.

4.2. Evaluating the Recovery Plan

In Table 5 the data for 2006 on the relevant variables and the optimum solution for the starting value of the trawl selectivity parameter are shown. As we can see, the real current situation differs hugely from the one referenced, therefore it is necessary to reduce the fishing effort exercised and catches, especially in the case of the trawl fleet.

Table 5. Comparison of the real situation with the optimal situation (γ_1 = 0.6; $\gamma_2 = 0.1$).

	X (tons)	E1 (fishing days)	E2 (fishing days)	h1 (tons)	h2 (tons)
2006	15.3	19.5*	21.4*	7.4	2.6
Optimal	36.1	3.1	18.9	2.8	8.3
Biological safety (B_{lim})	25.0				

1: trawl; 2: artisanal. Data in thousands. * Data corresponding to 2004.

The European Commission approved a recovery plan for this fish population in 2005 (EC, 2005). The aim of this plan is to place the stock within the limits of biological safety (Blim)[7] in ten years. For it, the plan establishes that the European Council will adopt decreasing TACs while the plan is in effect, but that decrease will never be able to be higher than 15% of the TACs for the preceding year (article 7).

On this basis, from the expression $X(t+1) = 8.9263\ \theta\ X(t)^{0.82310}$ - TAC(t) and from the 2006 TAC (agreed at 6,661 tons) it is possible to determine the number of years necessary to reach the biological safety limits and too the equilibrium stock value. Specifically, it would take between six and fourteen years to reach the Blim level and between eleven and eighteen years to reach the stationary biomass level, depending on the parameter γ_l (see Table 6).

[7] It was estimated at 25,000 tonnes (ICES).

Table 6. Approximation trend in accordance with TAC reduction (for different γ_1 and $\gamma_2 = 0.1$ values)

	15% TAC reduction					25% TAC reduction				
γ_1 / Years	0.6	0.5	0.4	0.3	0.2	0.6	0.5	0.4	0.3	0.2
X(1)	12,255	12,826	13,417	14,131	14,234	12,255	12,826	13,417	14,131	14,234
X(2)	10,356	11,469	12,655	14,134	14,351	11,022	12,135	13,321	14,800	15,018
X(3)	9,131	10,812	12,643	14,987	15,336	10,931	12,621	14,212	16,817	17,169
X(4)	8,482	10,793	13,352	16,687	17,189	11,769	14,095	16,208	20,035	20,541
X(5)	8,354	11,386	14,766	19,221	19,898	13,385	16,407	19,151	24,277	24,959
X(6)	8,729	12,575	16,865	22,544	23,411	15,642	19,399	22,864	29,324	30,193
X(7)	9,602	14,343	19,598	26,565	27,630	18,395	22,896	27,140	34,917	35,978
X(8)	10,968	16,646	22,885	31,147	23,411	21,486	26,712	31,762	*40,791*	*42,043*
X(9)	12,804	19,416	26,612	36,124	37,582	24,760	30,668	*36,519*	*46,705*	*48,139*
X(10)	15,063	22,556	30,642	*41,321*	*42,958*	28,075	34,607	*41,228*		
X(11)	17,670	25,952	34,835	*46,566*	48,367	31,314	*38,404*			
X(12)	20,532	29,485	*39,056*			34,387				
X(13)	23,689	33,041	43,188			*37,288*				
X(14)	26,869	*36,523*								
X(15)	29,982	39,853								
X(16)	32,956									
X(17)	35,738									
X(18)	*38,295*									

Source: Own compilation from $X(t+1) = 8.9263\ \theta\ X(t)^{0.82310}$ - TAC(t) and from TAC in 2006.

It is only possible to reach the safety limits for this species within the term established by the EU if trawl selectivity improves for values the same or lower than 0.4. On the contrary, if this fleet does not improve its level of selectivity, and if the aim is to be reached within a period of ten years, it would be necessary to introduce a greater drop in the TACs. Bearing in mind the model's parameters, the progressive reduction of the total catches is estimated at around 25% of the preceding TAC. This option would imply a period of approximately nine and thirteen years to reach the Blim level and the stationary resource level (Table 6). Once these levels have been reached the TAC could be increased to the values shown in Table 5.

5. Discussion

Since the European hake stock is overexploited and presents reduced reproductive capacity, the European Commission (EC) drew up a Recovery Plan approved by the Council in October 2005 and which came into effect in 2006, which aims to return this species within biological safety limits in ten years (EC, 2005). The recovery plan includes the establishment of decreasing Total Allowable Catches (TACs) for the southern hake stock, the reduction of total effort in the fishery and the need for that one of the fleets operating in this fishery (trawling fleet) improve its degree of selectivity by concentrating on individuals of a larger size.

In this chapter, we evaluate this management plan, put forward different scenarios and analyse how levels of biomass and catches equilibrium would be affected. A model for considering the size-selectivity of the fishing gears in the dynamic of resource is proposed for evaluating the recovery plan of hake fishery in VIIIc and IXa ICES areas. For it, we use a selectivity indicator which makes it possible to discover the impact of each technology on the marine resource. The results obtained show that as the selectivity of the fleet which is less conservationist with the marine resource improves, the optimum level of the natural resource increases, while the global effort level drops.

The starting situation in the hake fishery is far from what would be considered socially optimal or sustainable, making it necessary to reduce the pressure on the marine resource, situate the species within biological safety limits and reach optimum levels. For this to occur, the selectivity of the fleet which is less conservationist with the marine resource must improve. If not, it would be necessary to introduce a greater reduction of the total annual catches authorised for this species than that contemplated in the recovery plan.

Finally, we should underline that the measures proposed are based on economic efficiency criteria. More collaboration between biologists and economists is necessary for defining the bioeconomic model and incorporating the age structure of the biomass.

REFERENCES

Armstrong C, Clark D (1997) Just fishing? Equity and Efficiency in Fisheries Management Regimes. *Marine Resource Economics* 12: 203-220.

Bjorndal T, Lindroos M (2003) *International Management of North Sea Herring* In abstracts of the 12th Annual Conference of European Association of Environmental and Resources Economists.

Boyce, J (1996) An Economic Analysis of the Fisheries Bycatch Problem. *Journal of Environmental Economics and Management* 31, 314-336.

Clark C (1990) Mathematical Bioeconomics. *The Optimal Management of Renewable Resources,* 2nd edn. Wiley-Interscience Publication, New York.

Costa Duarte C, Brasao A, Pintasilgo P (2000) Management of the Northern Atlantic Bluefin Tuna: An application of C-Games. *Marine Resources Economics* 14: 21-36.

Cullen R (2000) Selection of Mechanisms to Achieve Internalisation of Fishing Externalities. In abstracts of the *X Conference of International Institute of Fisheries Economics and Trade*, http://osu.orst.edu/dept/IIFET/2000.

Economic Assessment of EU Fisheries (1998-2006), Economic Performance of Selected European Fishing Fleets, Annual Reports. European Communities, Brussels.

Escapa M, Prellezo R (2003) Fishing Technology and Optimal Distribution of Harvest Rates. *Environmental and Resource Economics* 25: 377-394.

European Commission (2005) Plan for recovery measures for southern hake, *COM* 818. European Communities, Brussels.

European Commission (2007) Communication from Commission on rights-based management tools in fisheries, *SEC* 247. European Communities, Brussels.

Gallastegui C (1983) An economic analysis of sardine in the Gulf of Valencia (Spain). *Journal of Environmental Economics and Management* 10: 138-150.

Garza-Gil M (1998) ITQ Systems in Multifleet Fisheries. An Application for Iberoatlantic Hake. *Environmental and Resource Economics* 15: 93-106.

Garza M.D. and Varela M. (2007) Bioeconomic management and fishing selectivity: An application to the European Hake fishery. *American Journal of Agricultural and Biological Sciences* 2 (2): 69-74.

ICES (several years) *Hake-Southern Stock Report*. www.ices.dk

Lleonart J, Recasens L (1996) Fisheries and Environment in the Mediterranean Sea. Resources and Environmental Issues relevant to Mediterranean Fisheries Management, *Studies and Reviews General Fisheries Council for the Mediterranean* N° 66. FAO, Rome.

Prellezo R, Gallastegui C, (2003) Gear Selectivity Based Regulation in a Mixed Fishery. *12th Annual Conference of European Association of Environmental and Resource Economists.*

Surís J (1993) Regulation of the Iberoatlantic Sardine Fishery. *Environmental and Resources Economic* 3: 457-470.

Trujillo V, Meixide M, Porteiro C, Pérez N, Pereiro F (1991) *Mesh size and effort changes in muiltispecies fisheries in ICES divisions VIIIc and IXa.* ICES C.M./G: 51.

Turner, M (1997) Quota Induced Discarding in Heterogeneous Fisheries. *Journal of Environmental Economics and Management* 33: 186-195.

Ward, J (1994) The Bioeconomic Implications of a Bycatch Reduction Device a Stock Conservation Management Measure. *Marine Resource Economics* 9: 227-240.

Weitzman M (2002) Landing Fee vs. Harvest Quotas with Uncertain Fish Stocks. *Journal of Environmental Economics and Management* 43: 325-338.

In: Fishery Management
Editor: Janice S. Intilli

ISBN: 978-1-61209-682-7

Chapter 10

COOPERATION MAINTENANCE IN FISHERY PROBLEMS

V.V. Mazalov*[*] and A.N. Rettieva [†]
Institute of Applied Mathematical Research,
Karelian Research Center of RAS
185910, Russia, Petrozavodsk, Pushkinskaya str., 11
phone: (8142)781108, fax: (8142)766313
URL: http://mathem.krc.karelia.ru

Abstract

The chapter is devoted to dynamic game models related to a bioresource management problem (fishery problems). The main goal of this work is to investigate the schemes to maintain the cooperative exploitation of fish resources.

We consider different game-theoretic models: discrete-time and continuous, with finite and infinite planning horizon, with different growth rules and players' payoffs functions.

We derive the Nash and the cooperative equilibria for considered models. Then we construct schemes to maintain the cooperation. In the context of cooperative incentive equilibrium we compare traditional method and our approach with changing harvesting territory. In the context of time-consistency we construct imputation set and time-consistent imputation distribution procedure.

Keywords: dynamic games, bioresource management problem, discrete-time games, Nash equilibrium, cooperative equilibrium, incentive equilibrium

AMS Subject Classification: 91A, 49N, 93C.

1. Introduction

Cooperation plays an important role in bioresource management problems because it causes minimal damage to the environment. Such cooperative behavior is possible if the consequences of non-cooperation will be taken into account by the rules of a dynamic game.

[*]E-mail address: vmazalov@krc.karelia.ru
[†]E-mail address: annaret@krc.karelia.ru

There are several methodological schemes to maintain the cooperation. Here we consider two ways: incentive equilibrium and time-consistent imputation distribution procedure.

The concept of cooperative incentive equilibrium was introduced in (Ehtamo & Hamalainen, 1993), as a natural extension of Osborn's work (Osborn, 1976) about cartel stability. In traditional concept players punish each other for a deviation from cooperative behavior by changing their optimal cooperative strategies. In the considered problems, the center punishes players for deviating from the cooperative equilibrium by changing the harvesting territory.

The question of dynamic stability in differential games has been investigated in the past three decades. Haurie (1976) raised the problem of instability of the Nash bargaining solution. The concept of time-consistency (dynamic stability) was introduced by Petrosjan (1977). Time-consistency involves the property that, as the cooperation develops, participants are guided by the same optimality principle at each time moment and hence do not have incentives to deviate from cooperation. Petrosjan (Petrosjan & Danilov, 1982) has developed the notion of time-consistent imputation distribution procedure. Petrosjan (1993) and Petrosjan & Zenkevich (1996) presented a method of regularization to construct time-consistent solutions. Petrosjan & Zaccour (2003) presented time-consistent Shapley value allocation in a differential game of pollution cost reduction. Yeung (2006) introduced the "irrational-behavior-proofness" condition that guaranties the stability of cooperative agreement against unpredictable collapse of the coalition.

In the chapter we consider different game-theoretic models: discrete-time and continuous, with finite and infinite planning horizon, with different growth rules and players' payoffs functions.

First, for considered models we derive the Nash and the cooperative equilibria. And the result shows that noncooperative behavior is inefficient due to overfishing.

Second, we construct schemes to maintain the cooperation. In the context of cooperative incentive equilibrium we compare traditional method and our approach with changing harvesting territory.

In the context of time-consistency we construct imputation set and time-consistent imputation distribution procedure. We introduce a new condition which stimulates the maintenance of cooperative agreement and compare it with "irrational-behavior-proofness" condition (Yeung, 2006). We present the results of numerical modelling and compare different considered approaches.

2. Main Concepts

2.1. Optimality Concepts

Differential or dynamic games study a class of decision problem, under which the evolution of the state is described by a differential equation and the players act throughout a time interval.

Typical control system can be described in terms of an ordinary differential equation of the following form:

$$x'(t) = f(x(t), u(t)), \; x(t_0) = x_0, \tag{2.1}$$

where $x = (x_1, \ldots, x_m) \in R^m$ and $u = (u_1, \ldots, u_n) \in U$ are the phase vector of the system and the vector of player's control, respectively. U – is the set of admissible controls and usually $U \in R^n$.

Here the time interval can be finite $[t_0, T]$ or infinite $[t_0, \infty)$.

Or in discrete-time

$$x_{t+1} = f(x_t, u_t), \quad x_{t_0} = x_0 . \tag{2.2}$$

Definition 2.1. *Dynamic game is*

$$\Gamma = < N, x, \{U_i\}_{i=1}^n, \{J_i\}_{i=1}^n > ,$$

where $N = \{1, \ldots, n\}$ *– set of players,* $U_1, \ldots, U_n$ *– the sets of player's admissible controls, and* $J_i(u)$ *– the payoff of player* $i \in N$.

The players' payoff functions can be presented in different ways. In the chapter we consider utility functionals in special forms which are usually used in economic problems:

$$J_i(u) = \int_{t_0}^{T} e^{-\rho t} g_i(x(t), u(t))dt + G_i(x(T)) \longrightarrow max ,$$

and

$$J_i(u) = \sum_{t=0}^{\infty} \delta^t g_i(x_t, u_t) \longrightarrow max ,$$

where $0 < \rho < 1$ and $0 < \delta < 1$ – the discount factors.

There are cooperative and non-cooperative dynamic games.

In non-cooperative game theory the main solution concept is the Nash equilibrium.

Definition 2.2. *The Nash equilibrium in game* Γ *is the set of controls (strategies)* $(u_1^*, \ldots, u_n^*)$ *such as*

$$J_i(u_1^*, \ldots, u_i, \ldots, u_n^*) \leq J_i(u_1^*, \ldots, u_i^*, \ldots, u_n^*)$$

for all arbitrary admissible u_i, $i = 1, \ldots, n$.

Under full cooperation the sum of players' total utilities is maximized:

$$J(u) = \sum_{i=1}^{n} J_i(u) = \int_{t_0}^{T} e^{-\rho t} \sum_{i=1}^{n} g_i(x(t), u(t))dt + \sum_{i=1}^{n} G_i(x(T)) \tag{2.3}$$

or in discrete time

$$J(u) = \sum_{i=1}^{n} J_i(u) = \sum_{t=0}^{\infty} \delta^t \sum_{i=1}^{n} g_i(x_t, u_t)dt . \tag{2.4}$$

Also players can form the coalitions less than grand coalition.

Denote the profit of coalition $S \in N$ as

$$J^S(u) = \sum_{i \in S} J_i(u) .$$

The solution optimality principle for a cooperative game includes:

1. an agreement on a set of cooperative controls,
2. a mechanism to distribute total payoff among players.

Then the characteristic function $V(S, 0)$ should be determined: $V(S, 0)$ is the profit of coalition S in the equilibrium, when all other players form a coalition $N \setminus S$ and play against coalition S.

How to distribute cooperative gain between players:

Definition 2.3. *A vector*

$$\xi = \{\xi(0) = (\xi_1(0), \ldots, \xi_n(0))$$

that satisfies the conditions:

1. $\sum_{i=1}^{n} \xi_i(0) = V(N, 0)$,

2. $\xi_i(0) \geq V(i, 0),\ i = 1, \ldots, n$

is called an imputation of the game Γ.

Analogously we determine the characteristic function $V(S, t)$ and the imputation set $\xi(t) = (\xi_1(t), \ldots, \xi_n(t))$ for every subgame started from the state $x(t)$ (x_t) at time moment t.

There are several cooperative optimality principles: proportional solution, C–core, n–core, the Shapley value or another.

Here we will use the Shapley value. The Shapley value is one of the most popular mechanism of the profit distribution in cooperative game theory. It gives an imputation rule for sharing the cooperative profit among the members in a coalition as:

$$\xi_i = \sum_{S \in N} \frac{(n - |S|)!(|S| - 1)!}{n!} \left[V_{\{S\}} - V_{\{S \setminus i\}}\right],\ i \in N = \{1, \ldots, n\},$$

where n – number of players, $|S|$ – number of players in coalition S, $V_{\{S\}}$ – profit of coalition S and $V_{\{S\}} - V_{\{S \setminus i\}}$ – marginal contribution of player i to the coalition S.

2.2. Cooperation Maintenance Concepts

Cooperation plays an important role in bioresource management problems because it causes minimal damage to the environment. Such cooperative behavior is possible if the consequences of non-cooperation will be taken into account by the rules of a dynamic game. There are several methodological schemes to maintain the cooperation. Here we consider two ways: incentive equilibrium and time-consistent imputation distribution procedure.

The concept of cooperative incentive equilibrium was introduced in (Ehtamo & Hamalainen, 1993), as natural extension of Osborn's work (Osborn, 1976) about cartel stability. In our problem statement the center punishes players for a deviation from cooperative equilibrium changing the harvesting territory.

This concept is used for the problem with two players. Following (Ehtamo & Hamalainen, 1993) we assume that the strategy of the player i is a causal mapping $\gamma_i : D_j \to D_i$ $(u_i \in D_i)$.

Definition 2.4. *A strategy pair* (γ_1, γ_2) *is called the cooperative incentive equilibrium for the problem (2.1), (2.3) (or (2.2), (2.4)) with two players if*

$$u_1^d = \gamma_1(u_2^d)\,, \quad u_2^d = \gamma_2(u_1^d)\,,$$

$$J_1(u_1^d, u_2^d) \geq J_1(u_1, \gamma_2(u_1))\,, \ \ \forall u_1(\cdot) \in D_1\,,$$

$$J_2(u_1^d, u_2^d) \geq J_2(\gamma_1(u_2), u_2)\,, \ \ \forall u_2(\cdot) \in D_2\,,$$

where u_1^d, u_2^d *– the cooperative equilibrium.*

The incentive equilibrium is applied for maintenance the cooperation and punishing the player who deviates. In traditional statement players control their behavior, punishing for deviation by changing the cooperative strategies. We consider the new scheme where the center controls the cooperation agreement.

The concept of time-consistency (dynamic stability) was introduced by L.A. Petrosjan (Petrosjan, 1977). Time-consistency involves the property that, as the cooperation develops participants are guided by the same optimality principle at each time moment and hence don't have incentives to deviate from cooperation. In the work (Petrosjan & Danilov, 1982) it was developed the notion of time-consistent imputation distribution procedure.

Definition 2.5. *The vector* $\beta(t) = (\beta_1(t), \ldots, \beta_n(t))$ *is an imputation distribution procedure (Petrosjan & Danilov, 1982; Petrosjan & Zaccour, 2003) if*

$$\xi_i(0) = \int_0^\infty e^{-\rho t}\beta_i(t)\,dt\,, i = 1, \ldots, n\,, \tag{2.5}$$

or for discrete-time problem

$$\xi_i(0) = \sum_{t=0}^{\infty} \delta^t \beta_i(t)\,, i = 1, \ldots, n\,. \tag{2.6}$$

The main idea of this scheme is to distribute the cooperation gain along the game path. Then $\beta_i(t)$ can be interpreted as the payment to player i at time moment t.

Definition 2.6. *The vector* $\beta(t) = (\beta_1(t), \ldots, \beta_n(t))$ *is a time-consistent imputation distribution procedure (Petrosjan & Zaccour, 2003) if*

$$\xi_i(0) = \int_0^t e^{-\rho\tau}\beta_i(\tau)\,d\tau + e^{-\rho t}\xi_i(t)\,, i = 1, \ldots, n\,, \tag{2.7}$$

or for discrete-time problem

$$\xi_i(0) = \sum_{\tau=0}^{t-1} \delta^\tau \beta_i(\tau) + \delta^t \xi_i(t)\,, i = 1, \ldots, n\,, \tag{2.8}$$

where $\xi_i(t)$ *is the imputation for player* i *at time* t.

Time consistency guarantees that the extension of the solution policy to a situation with a later starting time and along the optimal trajectory remains optimal.

Definition 2.7. *The imputation* $\xi = (\xi_1, \dots, \xi_n)$ *satisfies "irrational-behavior-proofness" condition (Yeung, 2006) if*

$$\int_0^t e^{-\rho\tau}\beta_i(\tau)\,d\tau + e^{-\rho t}V(i,t) \geq V(i,0)\,,$$

or for discrete-time problem

$$\sum_{\tau=0}^{t} \delta^\tau \beta_i(\tau) + \delta^{t+1}V(i,t+1) \geq V(i,0)$$

for all $t \geq 0$, *where* $\beta(t) = (\beta_1(t), \dots, \beta_n(t))$ – *time-consistent imputation distribution procedure.*

If this condition is satisfied, then player i is irrational-behavior-proof because irrational actions that break the cooperative agreement will not bring his payoff below his initial noncooperative payoff.

We introduce a new condition which is stronger than the Yeung's condition and is easier to verify.

Definition 2.8. *The imputation* $\xi = (\xi_1, \dots, \xi_n)$ *satisfies each step rational behavior condition if*

$$\beta_i(t) + \delta V(i,t+1) \geq V(i,t)$$

for all $t \geq 0$, *where* $\beta(t) = (\beta_1(t), \dots, \beta_n(t))$ – *time-consistent imputation distribution procedure.*

The proposed condition offers an incentive to player i to maintain cooperation because at every step she gains more from cooperation than from noncooperative behavior.

3. Continuous-Time Models

3.1. Bioresource Management Problem with Finite Horizon

The dynamic game model related to a bioresource management problem (fish catching) is considered. The center (referee) shares a reservoir between the competitors. The players (countries) which harvest the fish stock are the participants of this game. Each player is independent decision maker being guided by maximization of the profit of fish sale.

In the series of papers (Mazalov & Rettieva, 2005, 2006, 2008) it was developed the new approach where the policy of the center is to determine the optimal share of the aquatic environment where fishing is prohibited. In the chapter we apply the author's approach to bioresource sharing problem for two players under cooperation.

For finite horizon the Pontryagin's maximal principle (Pontryagin, Boltyanskii and al., 1976) is applied to determine the equilibrium. The incentive equilibria are constructed in the case when the players punish each other for a deviation from the cooperative equilibrium, and in the case when the center punishes them for a deviation.

Let us divide the water area into two parts, s and $1-s$, where two countries exploit the fish stock. The center (referee) shares the reservoir. Players (countries) exploit the fish stock during T time periods on their territory.

The dynamics of the fishery is described by the equation

$$x'(t) = F(x(t)) - q_1 E_1(t)(1-s)x(t) - q_2 E_2(t) s x(t)\,, \quad 0 \le t \le T,\ x(0) = x_0\,, \quad (3.1)$$

where $x(t) \ge 0$ – size of the population at time t; F – natural growth function of the population; $E_1(t), E_2(t) \ge 0$ – countries' fishing efforts measured as the number of vessels involved in fishing at time t and $q_1, q_2 > 0$ – catchability coefficients related to the unit fishing effort of the country. The condition on controls to be integer is ignored in our problem.

We assume that $E_1,\ E_2$ belong to decision sets $D_1,\ D_2$. Let $D_1 = D_2 \subseteq C([0,\infty))$.

Assume that fish population grows according to Verhulst model

$$F(x) = rx\Big(1 - \frac{x}{K}\Big)\,,$$

where $r > 0$ – the intrinsic growth rate, and $K > 0$ – maximal natural object capacity.

The players' net revenues over a fixed time period $[0,T]$ are

$$J_1 = g_1(x(T)) + \int_0^T e^{-\rho_1 t}[q_1 E_1(t)(1-s)x(t)(p_1 - k_1 q_1 E_1(t)(1-s)x(t))]dt\,,$$
$$J_2 = g_2(x(T)) + \int_0^T e^{-\rho_2 t}[q_2 E_2(t) s x(t)(p_2 - k_2 q_2 E_2(t) s x(t))]dt\,,$$

where p_i – the price, k_i – catching costs, ρ_i – the discount factor, $i = 1,\ 2$.

Functions $g_i(x)$ describe the salvage value of the stock at time T. Following usual assumptions on utility function we suppose that $g_i'(x) \ge 0,\ g_i''(x) \le 0\,, i = 1,\ 2$.

The player's profit here is presented as income on time period $[0,T]$ which depends on the difference between the price and the catching costs with discounting. Here catching costs have quadratic form.

Substituting $a_i = 2k_i q_i^2 e^{-\rho_i t}$, $b_i = p_i q_i e^{-\rho_i t}$, we receive the profits of the form

$$J_1 = g_1(x(T)) + \int_0^T [-\tfrac{1}{2} a_1 E_1^2(t)(1-s)^2 x^2(t) + b_1 E_1(t)(1-s)x(t)]dt\,,$$
$$J_2 = g_2(x(T)) + \int_0^T [-\tfrac{1}{2} a_2 E_2^2(t) s^2 x^2(t) + b_2 E_2(t) s x(t)]dt\,. \quad (3.2)$$

We investigate different types of the equilibrium for considered dynamic game (3.1)-(3.2).

Ecological equilibrium. Let's investigate the asymptotic solution of our problem. Assume that the players wish to maximize their profits (expressions in the integrals (3.2)). So, we should solve the problem

$$J_1^0(E_1) = -\frac{1}{2} a_1 E_1^2 (1-s)^2 x^2 + b_1 E_1 (1-s)x,$$
$$J_2^0(E_2) = -\frac{1}{2} a_2 E_2^2 s^2 x^2 + b_2 E_2 s x\,,$$
$$J_1^0(E_1) \Longrightarrow \max_{E_1}\,,\ J_2^0(E_2) \Longrightarrow \max_{E_2}\,.$$

We receive the steady–state solution

$$\bar{E}_1 = \frac{p_1}{2k_1 q_1 (1-s)\bar{x}}\,, \quad \bar{E}_2 = \frac{p_2}{2k_2 q_2 s\bar{x}}\,.$$

Asymptotic size of population

$$\bar{x} = \frac{K}{r}(r - q_1 E_1(1-s) - q_2 E_2 s) \tag{3.3}$$

we received setting the right-hand side of (3.1) equal to 0.

Finally, substituting $\bar{E}_1$, $\bar{E}_2$ into (3.3) we obtain the ecological equilibrium

$$\bar{x} = \frac{K}{2} + \frac{1}{2}\sqrt{K^2 - \frac{2Kp_1}{rk_1} - \frac{2Kp_2}{rk_2}}\,.$$

Cooperative equilibrium. We solve the following optimization problem to define the cooperative equilibrium.

$$\begin{cases} \max(\mu_1 J_1(E_1, E_2) + \mu_2 J_2(E_1, E_2))\,, \\ \quad \text{where } x(t) \text{ is defined in (3.1)}\,, \end{cases} \tag{3.4}$$

where $\mu_1, \mu_2 > 0$ – weighting coefficients which reflect significance of the players, $\mu_1 + \mu_2 = 1$. The solution of this problem is given in the next theorem.

Theorem 3.1. *The cooperative equilibrium of the problem (3.1), (3.4) is*

$$E_1^d(t) = \frac{b_1 - \mu_1^{-1} q_1 \lambda(t)}{a_1(1-s)x(t)}\,, \quad E_2^d(t) = \frac{b_2 - \mu_2^{-1} q_2 \lambda(t)}{a_2 s x(t)}\,, \tag{3.5}$$

with $\lambda(t)$ *satisfying the system of differential equations*

$$\begin{cases} \lambda'(t) = -\lambda(t)\Big(r - \dfrac{2rx(t)}{K}\Big)\,, \ \lambda(T) = \mu_1 g_1'(x(T)) + \mu_2 g_2'(x(T))\,, \\ x'(t) = rx(t)\Big(1 - \dfrac{x(t)}{K}\Big) - \displaystyle\sum_{i=1}^{2} \frac{q_i}{a_i}(b_i - \mu_i^{-1} q_i \lambda(t))\,; \ x(0) = x_0\,. \end{cases}$$

Proof. Let's use the maximal principle (Pontryagin, Boltyanskii and al., 1976). The Hamilton function is

$$\begin{aligned} H(E_1, E_2, s, x) &= \mu_1(-\tfrac{1}{2}a_1 E_1^2 (1-s)^2 x^2 + b_1 E_1 (1-s)x) + \\ &+ \mu_2(-\tfrac{1}{2}a_2 E_2^2 s^2 x^2 + b_2 E_2 s x) + \lambda(rx - \frac{rx^2}{K} - q_1 E_1(1-s)x - q_2 E_2 s x)\,. \end{aligned}$$

Solving the system

$$\begin{cases} \dfrac{\partial H}{\partial E_1} = 0\,, \\ \dfrac{\partial H}{\partial E_2} = 0 \end{cases}$$

we receive that the optimal controls are of the form

$$E_1^d = \frac{b_1 - \mu_1^{-1} q_1 \lambda}{a_1(1-s)x}\,, \quad E_2^d = \frac{b_2 - \mu_2^{-1} q_2 \lambda}{a_2 s x}\,,$$

where $\lambda(t)$ – conjugate variable satisfying the equation

$$\lambda'(t) = -\frac{\partial H}{\partial x}, \ \lambda(T) = \sum_{i=1}^{2} \mu_i g_i'(x(T)) .$$

Let's prove the optimality of the controls. Substituting the expressions $E_1^d(t)$ and $E_2^d(t)$ in $\frac{\partial H}{\partial x}$ we obtain that the equation for the conjugate variable takes the form

$$\lambda'(t) = -\lambda(t)(r - \frac{2rx(t)}{K}), \ \lambda(T) = \mu_1 g_1'(x(T)) + \mu_2 g_2'(x(T)).$$

Then

$$\lambda(t) = \sum_{i=1}^{2} \mu_i g_i'(x(T)) e^{\int_t^T (r - 2rx(\tau)/K)\, d\tau} > 0 .$$

Let's denote

$$H^0(x, \lambda, t) = \max_{E_1 \in D_1, E_2 \in D_2} H(x, E_1, E_2, \lambda, t) .$$

Substituting $E_1^d(t)$ and $E_2^d(t)$ and simplifying we obtain

$$H^0(x, \lambda, t) = \lambda F(x) + \frac{(\mu_2 b_2 - \lambda q_2)^2}{2a_2 \mu_2} + \frac{(\mu_1 b_1 - \lambda q_1)^2}{2a_1 \mu_1} .$$

$$H^0_{xx}(x, \lambda, t) = -\frac{2r}{K}\lambda .$$

With $\lambda(t) > 0$ it yields that H^0 – concave on x.
Using concavity and definition of H^0 we can write

$$H(x, \lambda, t) \leq H^0(x^d, \lambda, t) - \lambda'(t)(x(t) - x^d(t)) .$$

Let's integrate this inequality on t from 0 to T

$$\int_0^T H(x, \lambda, t) \leq \int_0^T H^0(x^d, \lambda, t) - \lambda(t)(x(t) - x^d(t))|_0^T + \int_0^T \lambda(t)(x'(t) - (x^d(t))')dt .$$

Let's denote $g(x(T)) = \mu_1 g_1(x(T)) + \mu_2 g_2(x(T))$, $J(E_1, E_2) = \mu_1 J_1(E_1, E_2) + \mu_2 J_2(E_1, E_2)$ and $J^d(E_1^d, E_2^d) = \mu_1 J_1(E_1^d, E_2^d) + \mu_2 J_2(E_1^d, E_2^d)$. Then we can rewrite the inequality as

$$\begin{aligned} &J(E_1, E_2) - g(x(T)) + \textstyle\int_0^T \lambda(t)(F(x(t)) - q_1 E_1(t)(1 - s)x(t) - q_2 E_2(t) s x(t))dt \leq \\ &\leq J^d(E_1^d, E_2^d) + \textstyle\int_0^T \lambda(t)(F(x^d(t)) - q_1 E_1^d(t)(1 - s)x^d(t) - q_2 E_2^d(t) s x^d(t))dt - \\ &-g(x^d(T)) - \lambda(T)(x(T) - x^d(T)) + \textstyle\int_0^T \lambda(t)x'(t)dt + \int_0^T \lambda(t)(x^d(t))'dt \end{aligned}$$

Simplifying we get

$$\begin{aligned} &J(E_1, E_2) \leq J^d(E_1^d, E_2^d) + g(x(T)) - g(x^d(T)) - \lambda(T)(x(T) - x^d(T)) \leq \\ &\leq J^d(E_1^d, E_2^d) - \lambda(T)(x(T) - x^d(T)) + g_x'(x^d(T))(x(T) - x^d(T)) . \end{aligned}$$

Note that $\lambda(T) = g'_x(x^d(T))$ we get

$$J(E_1, E_2) \leq J^d(E_1^d, E_2^d)$$

or

$$\mu_1 J_1(E_1(t), E_2(t)) + \mu_2 J_2(E_1(t), E_2(t)) \leq \mu_1 J_1(E_1^d(t), E_2^d(t)) + \mu_2 J_2(E_1^d(t), E_2^d(t)) .$$

So we proved that $E_1^d(t)$ and $E_2^d(t)$ – solution of the problem (3.4).

Incentive equilibrium. Following (Ehtamo & Hamalainen, 1993) we assume that the strategy of the player i is a causal mapping $\gamma_i : D_j \to D_i$ $(E_i \in D_i)$.

Assume that the players punish each other for a deviation from the cooperative equilibrium by increasing the control on the value which is proportional to the difference $E_i(t) - E_i^d(t)$. Then, following (Ehtamo & Hamalainen, 1993), it is not difficult to prove the next statement.

Proposition 3.1. *The incentive equilibrium of the problem (3.1), (3.4) is*

$$\gamma_1(E_2(t)) = E_1^d(t) + \eta_1(t)(E_2(t) - E_2^d(t)) , \quad \gamma_2(E_1(t)) = E_2^d(t) + \eta_2(t)(E_1(t) - E_1^d(t)) ,$$

where

$$\eta_1(t) = \frac{q_2 \mu_1 \lambda_1(t) s}{q_1 \mu_2 \lambda_2(t)(1-s)} , \qquad \eta_2(t) = \frac{1}{\eta_1(t)} ,$$

$E_1^d(t)$, $E_2^d(t)$ are defined in (3.5) when $\lambda = \mu_1\lambda_1 + \mu_2\lambda_2$, and conjugate variables $\lambda_i(t)$, $i = 1, 2$, satisfy the next system of equation:

$$\begin{cases} \lambda_1'(t) = -\lambda_1(t)\Big(r - \dfrac{2rx(t)}{K}\Big) + \lambda_1(t) q_2 s (E_2^d(t) - \eta_2 E_1^d(t)) , \\ \lambda_2'(t) = -\lambda_2(t)\Big(r - \dfrac{2rx(t)}{K}\Big) + \lambda_2(t) q_1 (1-s)(E_1^d(t) - \eta_1 E_2^d(t)) , \\ x'(t) = rx(t)\Big(1 - \dfrac{x(t)}{K}\Big) - q_1 E_1^d(t)(1-s)x(t) - q_2 E_2^d(t) s x(t) , \\ \lambda_1(T) = \mu_1 g_1'(x(T)), \ \ \lambda_2(T) = \mu_2 g_2'(x(T)), \ \ x(0) = x_0 . \end{cases}$$

Denote s^d – territory sharing under cooperation. Let's change the scheme of the punishment for a deviation from the cooperative equilibrium. We assume that the players deviating from the cooperative equilibrium point are punished by the center rather than by themselves, as was in (Ehtamo & Hamalainen, 1993). Assume that, if the first player deviates the center increases s, and if the second player deviates – decreases s. Consider the second player's deviation

$$E_2(t) = E_2^d(t) + \Delta .$$

The center's strategy is sought in the form

$$s^*(t) = s^d - \eta(t)(E_2(t) - E_2^d(t)) .$$

Then according to Theorem 3.1 the first player's punishment strategy is

$$\gamma_1(E_2(t)) = \frac{b_1 - \mu_1^{-1} q_1 \lambda(t)}{a_1(1 - s^d + \eta(t)(E_2(t) - E_2^d(t)))x(t)} .$$

Next we should determine the coefficient $\eta(t)$. For that we consider the next problem

$$\begin{cases} & \max(J_2(\gamma_1(E_2(t)), E_2(t)))\,, \\ x'(t) = & rx(t)(1-\frac{x(t)}{K}) - q_1\gamma_1(E_2(t))(1-s^d+\eta(t)(E_2(t)-E_2^d(t)))x(t) - \\ - & q_2E_2(t)(s^d-\eta(t)(E_2(t)-E_2^d(t)))x(t)\,. \end{cases} \tag{3.6}$$

Using the Pontryagin's maximal principle we write the Hamilton function

$$H_2(E_2, s^*, x) = b_2E_2(s^d-\eta(E_2-E_2^d))x - \frac{1}{2}a_2E_2^d(s^d-\eta(E_2-E_2^d))^2x^2 +$$

$$+\lambda_2(rx - \frac{rx^2}{K} - q_2E_2(s^d-\eta(E_2-E_2^d))x - q_1\gamma_1(E_2)(1-s^d+\eta(E_2-E_2^d))x) =$$

$$= -\frac{1}{2}a_2E_2^2(s^d-\eta(E_2-E_2^d))^2x^2 + (b_2-q_2\lambda_2)E_2(s^d-\eta(E_2-E_2^d))x +$$

$$+\lambda_2\Big(rx - \frac{rx^2}{K} - q_1\frac{b_1-\mu_1^{-1}q_1\lambda}{a_1}\Big).$$

The maximum of this function is achieved at

$$\frac{\partial H_2}{\partial E_2} = x(-a_2E_2(s^d-\eta(E_2-E_2^d))x + b_2 - \lambda_2 q_2)\,(s^d-\eta(E_2-E_2^d)-\eta E_2) = 0\,. \tag{3.7}$$

γ_1 is the incentive equilibrium if the cooperative equilibrium is the solution of the problem (3.6). Consequently, $E_2 = E_2^d$ should satisfy the equation (3.7). Here the expression in the first brackets is non-zero, and the expression in the second brackets is equal to zero when

$$\eta(t) = \frac{s^d}{E_2^d(t)}\,.$$

Analogously, when the first player deviates we define the center's strategy

$$s^* = s^d + \theta(t)(E_1(t) - E_1^d(t))\,,$$

and the second player's strategy

$$\gamma_2(E_1(t)) = \frac{b_2 - \mu_2^{-1}q_2\lambda(t)}{a_2(s^d+\theta(t)(E_1(t)-E_1^d(t)))x(t)}\,.$$

It yields

$$\theta(t) = \frac{1-s^d}{E_1^d(t)}\,.$$

So, we proved the next theorem.

Theorem 3.2. *The incentive equilibrium of the problem (3.1), (3.4) is*

$$\gamma_1(E_2(t)) = \frac{b_1-\mu_1^{-1}q_1\lambda(t)}{a_1(1-s_2^*(t))x(t)}\,, \quad \gamma_2(E_1(t)) = \frac{b_2-\mu_2^{-1}q_2\lambda(t)}{a_2s_1^*(t)x(t)}\,,$$

where

$$s_2^*(t) = s^d - \frac{s^d}{E_2^d(t)}(E_2(t)-E_2^d(t))\,, \quad s_1^*(t) = s^d + \frac{1-s^d}{E_1^d(t)}(E_1(t)-E_1^d(t))\,,$$

and $E_1^d(t)$, $E_2^d(t)$, $x(t)$, $\lambda(t)$ satisfy Theorem 3.1.

3.2. Continuous-Time Model with Linear Growth Function

Let's consider another continuous-time problem of bioresource management problem. As above we divide the water area into two parts, s and $1-s$, where two countries exploit the fish stock. The center (referee) shares the reservoir. Players (countries) exploit the fish stock during infinite time on their territory.

The dynamics of the fishery is

$$x'(t) = \varepsilon x(t) - q_1 E_1(t)(1-s)x(t) - q_2 E_2(t) s x(t)\,, \quad t \geq 0,\ x(0) = x_0\,, \tag{3.8}$$

where $x(t) \geq 0$ – size of the population at time t; ε – the natural growth coefficient; $E_1(t), E_2(t) \geq 0$ – countries' fishing efforts measured as the number of vessels involved in fishing at time t and $q_1, q_2 > 0$ – catchability coefficients related to the unit fishing effort of the country.

Assume that $E_1,\ E_2$ belongs to $D_1,\ D_2$. Let $D_1 = D_2 \subseteq C([0,\infty))$.

The players wish to minimize the next functionals

$$J_1 = \int\limits_0^\infty e^{-\rho t}[q_1 E_1(t)(1-s)x(t)(k_1 q_1 E_1(t)(1-s)x(t) - p_1)]dt\,,$$
$$J_2 = \int\limits_0^\infty e^{-\rho t}[q_2 E_2(t) s x(t)(k_2 q_2 E_2(t) s x(t) - p_2)]dt\,,$$

where p_i – the price of fish, k_i – catching costs, $i = 1, 2$.

Or substituting $a_i = 2k_i q_i^2$, $b_i = p_i q_i$, we get

$$\begin{aligned} J_1 &= \int\limits_0^\infty e^{-\rho t}[\tfrac{1}{2}a_1 E_1^2(t)(1-s)^2 x^2(t) - b_1 E_1(t)(1-s)x(t)]dt\,,\\ J_2 &= \int\limits_0^\infty e^{-\rho t}[\tfrac{1}{2}a_2 E_2^2(t) s^2 x^2(t) - b_2 E_2(t) s x(t)]dt\,. \end{aligned} \tag{3.9}$$

Again we find the cooperative equilibrium.

Theorem 3.3. *The cooperative equilibrium of the problem (3.8)-(3.9) is*

$$E_1^d(t) = \frac{2Ax(t) + p_1\mu_1 + B}{2k_1 q_1 (1-s)x(t)}\,, \quad E_2^d(t) = \frac{2Ax(t) + p_2\mu_2 + B}{2k_2 q_2 s x(t)}\,, \tag{3.10}$$

where

$$A = \frac{(2\varepsilon - \rho)\mu_1\mu_2 k_1 k_2}{\mu_1 k_1 + \mu_2 k_2}\,, \quad B = -\frac{A(p_1 k_2 + p_2 k_1)}{\varepsilon k_1 k_2}\,.$$

Proof. We use the Bellman principle. Let's consider the value function

$$V(x) = \min_{E_1, E_2} \int_0^\infty e^{-\rho t}[\mu_1 J_1(E_1, E_2) + \mu_2 J_2(E_1, E_2)]\,dt\,.$$

The Hamilton-Jacobi-Bellman equation has the form

$$\begin{aligned} \rho V(x) &= \mu_1(\tfrac{1}{2}a_1 E_1^2(1-s)^2 x^2 - b_1 E_1(1-s)x) + \mu_2(\tfrac{1}{2}a_2 E_2^2 s^2 x^2 - b_2 E_2 s x) + \\ &+ \tfrac{dV}{dx}(\varepsilon x - q_1 E_1(1-s)x - q_2 E_2 s x)\,. \end{aligned} \tag{3.11}$$

Minimizing on E_1, E_2 we get

$$E_1^d = \frac{\mu_1 b_1 + q_1 \frac{dV}{dx}}{\mu_1 a_1 (1-s)x}, \quad E_1^d = \frac{\mu_2 b_2 + q_2 \frac{dV}{dx}}{\mu_2 a_2 s x}. \tag{3.12}$$

Substituting the optimal controls in (3.11) we receive

$$\rho V(x) = -\frac{(\mu_1 b_1 + q_1 \frac{dV}{dx})^2}{2\mu_1 a_1} - \frac{(\mu_2 b_2 + q_2 \frac{dV}{dx})^2}{2\mu_2 a_2} + \frac{dV}{dx}\varepsilon x. \tag{3.13}$$

The value function sought in the quadratic form

$$V(x) = Ax^2 + Bx + C.$$

Substituting in (3.13) we get the equations for coefficients

$$\begin{aligned}
\rho A &= -\frac{2A^2 q_1^2}{\mu_1 a_1} - \frac{2A^2 q_2^2}{\mu_2 a_2} + 2A\varepsilon, \\
\rho B &= -\frac{2Aq_1(\mu_1 b_1 + Bq_1)}{\mu_1 a_1} - \frac{2Aq_2(\mu_2 b_2 + Bq_2)}{\mu_2 a_2} + \varepsilon B, \\
\rho C &= -\frac{(\mu_1 b_1 + Bq_1)^2}{2\mu_1 a_1} - \frac{(\mu_2 b_2 + Bq_2)^2}{2\mu_2 a_2},
\end{aligned}$$

which yields

$$A = \frac{(2\varepsilon - \rho)\mu_1 \mu_2 k_1 k_2}{\mu_1 k_1 + \mu_2 k_2}, \quad B = -\frac{(2\varepsilon - \rho)\mu_1 \mu_2 (p_1 k_2 + p_2 k_1)}{\varepsilon(\mu_1 k_1 + \mu_2 k_2)}.$$

Substituting it into (3.12) we get the optimal controls (3.10).

The dynamics under cooperative behavior becomes

$$x'(t) = \varepsilon x(t) - q_1 E_1^d(t)(1-s)x(t) - q_2 E_2^d(t) s x(t) = (\rho - \varepsilon)x(t) + \frac{(\varepsilon - \rho)(p_2 k_1 + p_1 k_2)}{2\varepsilon k_1 k_2},$$

which yields

$$x^d(t) = \frac{p_2 k_1 + p_1 k_2}{2\varepsilon k_1 k_2} + (x_0 - \frac{p_2 k_1 + p_1 k_2}{2\varepsilon k_1 k_2})e^{(\rho-\varepsilon)t}.$$

Cooperative incentive equilibrium. As above the strategy of the player i is a causal mapping $\gamma_i : D_j \to D_i$ ($E_i \in D_i$).

Denote s^d – territory sharing under cooperation. We assume that the center punish the players for deviating from the cooperative equilibrium changing the harvesting territory. If the first player deviates the center increases s^d, and if the second player deviates – decreases s^d.

Consider the second player's deviation

$$E_2 = E_2^d + \Delta.$$

The center's strategy is sought in the form

$$s^*(t) = s^d - \eta(t)(E_2(t) - E_2^d(t)).$$

Then according to Theorem 3.3 the first player's punishment strategy is

$$\gamma_1(E_2) = \frac{2Ax(t) + \mu_1 p_1 + B}{2k_1 q_1(1 - s^d + \eta(t)(E_2(t) - E_2^d(t)))x(t)} .$$

To determine the coefficient $\eta(t)$ we solve the next problem

$$\begin{cases} & \max(J_2(\gamma_1(E_2), E_2)) , \\ x'(t) &= \varepsilon x(t) - q_1\gamma_1(E_2)(1 - s^d + \eta(E_2 - E_2^d))x - \\ &- q_2 E_2(s^d - \eta(E_2 - E_2^d))x . \end{cases} \quad (3.14)$$

We solve this problem analogously to Theorem 3.3 and construct the Hamilton-Jacobi-Bellman equation. Maximum of the value function is achieved when

$$x(a_2 E_2(s^d - \eta(E_2 - E_2^d))x - b_2 + \frac{dV}{dx} q_2)\,(s^d - \eta(E_2 - E_2^d) - \eta E_2) = 0 . \quad (3.15)$$

γ_1 is the incentive equilibrium if the cooperative equilibrium is the solution of the problem (3.14). Consequently, $E_2 = E_2^d$ should satisfy the equation (3.15). Here the expression in the first brackets is non-zero, and the expression in the second brackets is equal to zero when

$$\eta(t) = \frac{s^d}{E_2^d(t)} .$$

Analogously, when the first player deviates we define the center's strategy

$$s^* = s^d + \theta(t)(E_1(t) - E_1^d(t)) ,$$

and the second player's strategy

$$\gamma_2(E_1) = \frac{2Ax(t) + \mu_2 p_2 + B}{2k_2 q_2(s^d + \theta(t)(E_1(t) - E_1^d(t)))x(t)} .$$

It yields

$$\theta(t) = \frac{1 - s^d}{E_1^d(t)} .$$

So, we proved the next theorem.

Theorem 3.4. *The cooperative incentive equilibrium of the problem (3.8)-(3.9) is*

$$\gamma_1(E_2) = \frac{2Ax(t) + B + \mu_1 p_1}{2k_1 q_1(1 - s_2^*(t))x(t)} , \quad \gamma_2(E_1) = \frac{2Ax(t) + B + \mu_2 p_2}{2k_2 q_2 s_1^*(t)x(t)} ,$$

where

$$s_2^*(t) = s^d - \frac{s^d}{E_2^d(t)}(E_2(t) - E_2^d(t)) , \quad s_1^*(t) = s^d + \frac{1 - s^d}{E_1^d(t)}(E_1(t) - E_1^d(t)) .$$

Time-consistent imputation distribution procedure. First we find the Shapley value.

Theorem 3.5. *The Shapley value of the problem (3.8)-(3.9) is*

$$(\xi_1, \xi_2)$$

where

$$\xi_i = \frac{1}{2}(A_{\xi_i}x^2 + B_{\xi_i}x + C_{\xi_1})\,,\ i = 1, 2\,, \tag{3.16}$$

$$A_{\xi_i} = \frac{2\varepsilon - \rho}{3}M_i\,,\ B_{\xi_i} = -\frac{2(2\varepsilon - \rho)}{3}M_iK\,,$$

$$C_{\xi_i} = \frac{k_ip_j^2\mu_i - k_jp_i^2(1+\mu_i)}{4k_1k_2\rho} + \frac{2\varepsilon - \rho}{3}M_iK\,, i = 1, 2\,,$$

$$M_i = k_i - k_j + \frac{3\mu_1\mu_2k_1k_2}{\mu_1k_1 + \mu_2k_2}\,,\ i, j = 1, 2,\ i \neq j\,,$$

$$K = \frac{p_1k_2 + p_2k_1}{2\varepsilon k_1k_2}\,.$$

Proof. To construct the Shapley value we consider all possible coalitions and find the profits. We use the Bellman principle analogously to Theorem 3.3.

We get the optimal controls and payoffs

1. Grand coalition $\{1, 2\}$

$$E_1^d = \frac{2Ax + p_1\mu_1 + B}{2k_1q_1(1-s)x}\,,\ E_2^d = \frac{2Ax + p_2\mu_2 + B}{2k_2q_2sx}\,,$$

$$V(x) = Ax^2 + Bx + C\,,$$

where

$$A = \frac{(2\varepsilon - \rho)\mu_1\mu_2k_1k_2}{\mu_1k_1 + \mu_2k_2}\,,\ B = -2AK\,,\ C = -\frac{\mu_1p_1^2k_2 + \mu_2p_2^2k_1}{4\rho k_1k_2} - \frac{1}{2}KB\,.$$

2. Coalition $\{1\}$

$$E_1 = \frac{2A_1x + p_1\mu_1 + B_1}{2k_1q_1(1-s)x}\,,$$

$$V_1(x) = A_1x^2 + B_1x + C_1\,,$$

where

$$A_1 = \frac{(2\varepsilon - \rho)k_1}{3}\,,\ B_1 = -\frac{2}{3}A_1K\,,\ C_1 = -\frac{p_1^2}{4k_1\rho} - \frac{1}{2}KB_1\,.$$

3. Coalition $\{2\}$

$$E_2 = \frac{2A_2x + p_2\mu_1 + B_2}{2k_2q_2sx}\,,$$

$$V_2(x) = A_2x^2 + B_2x + C_2\,,$$

where

$$A_2 = \frac{(2\varepsilon - \rho)k_2}{3}\,,\ B_2 = -\frac{2}{3}A_2K\,,\ C_2 = -\frac{p_2^2}{4k_2\rho} - \frac{1}{2}KB_2\,.$$

The Shapley value becomes

$$\xi_1 = \frac{1}{2}(A_1 + A - A_2)x^2 + \frac{1}{2}(B_1 + B - B_2)x + \frac{1}{2}(C_1 + C - C_2)\,,$$

$$\xi_2 = \frac{1}{2}(A_2 + A - A_1)x^2 + \frac{1}{2}(B_2 + B - B_1)x + \frac{1}{2}(C_2 + C - C_1)\,.$$

Substituting the coefficients we get the Shapley value in the form as it is presented in the Theorem 3.5.

Next we construct time-consistent imputation distribution procedure.

Theorem 3.6. *The vector* $\beta(t) = (\beta_1(t), \beta_2(t))$*, where*

$$\beta_i(t) = \frac{1}{2}A_{\xi_i}(2\varepsilon - \rho)x(t)^2 + A_{\xi_i}K(\rho - 2\varepsilon)x(t) + \frac{1}{2}\rho C_{\xi_i} - (\rho - \varepsilon)K^2 A_{\xi_i}\,,$$

$$x(t) = K + (x_0 - K)e^{(\rho-\varepsilon)t}\,,\ \ K = \frac{p_1 k_2 + p_2 k_1}{2\varepsilon k_1 k_2}$$

is time-consistent imputation distribution procedure.

Proof. Consider the vector $\beta(t)$ as

$$\beta_i(t) = \rho\xi_i(t) - \frac{d}{dt}\xi_i(t)\,.$$

Let's prove that this vector is imputation distribution procedure, from (2.5) we need to check

$$\int_0^\infty e^{-\rho t}\beta_i(t)\,dt = \int_0^\infty e^{-\rho t}(\rho\xi_i(t) - \frac{d}{dt}\xi_i(t))\,dt = -e^{-\rho t}\xi_i(t)|_0^\infty = \xi_i(0)\,.$$

Let's prove that this vector is time-consistent imputation distribution procedure, from (2.7), we consider

$$\begin{aligned}&\textstyle\int_0^t e^{-\rho\tau}\beta_i(\tau)\,d\tau + e^{-\rho t}\xi_i(t) = \int_0^t e^{-\rho\tau}(\rho\xi_i(\tau) - \frac{d}{dt}\xi_i(\tau))\,d\tau + e^{-\rho t}\xi_i(t) = \\ &= -e^{-\rho\tau}\xi_i(\tau)|_0^t + e^{-\rho t}\xi_i(t) = \xi_i(0)\,.\end{aligned}$$

Using (3.16) we get

$$\beta_i(t) = A_{\xi_i}(\frac{1}{2}\rho x(t) - x'(t))x(t) + \frac{1}{2}B_{\xi_i}(\rho x(t) - x'(t)) + \frac{1}{2}\rho C_{\xi_i}\,. \qquad (3.17)$$

The dynamics under cooperation becomes

$$x'(t) = (\rho - \varepsilon)(x(t) - K)\,.$$

Substituting it into (3.17) and simplifying we get the time-consistent imputation procedure as it is presented in Theorem 3.6.

3.3. Modelling

We use the long-term data of the white-fish population in Karelian lake – Syamozero (Sterligova et al., 2002). To estimate the size of population we use matrix method of stock estimation (Abakumov et al., 1994). We evaluate the parameters of the growth function using constructed estimations of population size. We consider the next growth function:

$$F(x) = \varepsilon x(t) + \alpha \,,$$

where ε – natural growth coefficient, α – minimum allowable size of population.

For this function the results we derived above are valid for

$$K = \frac{p_1 k_2 + p_2 k_1 - 2\alpha k_1 k_2}{2\varepsilon k_1 k_2} \,.$$

In Syamozero lake we can extract four regions where the population is exploited. According to long-term data we can say that in the first region there are 16 % of population, in the second – 25 %, in the third – 11% and in the forth – 48%. For modelling we join three first regions in one and we receive two regions with almost equal population size.

We consider the next set of parameters

$$\begin{aligned} \varepsilon &= -0.1 \,, & \alpha &= 110 \,, \\ \rho &= 0.02 \,, & p_1 = p_2 &= 6000 \,, \\ s^d &= 0.5 \,, & q_1 = q_2 &= 0.02 \,, \\ \mu_1 &= 0.5 \,, & \mu_2 &= 0.45 \,. \end{aligned}$$

Incentive equilibrium. We consider the case when after the deviation the player returns back to cooperation. Let $k_1 = k_2 = 2$. The initial size of the population is $x(0) = 5000$. The time moment when the second player deviates is $t_0 = 10$ and the size of a deviation is $\Delta = 0.5$.

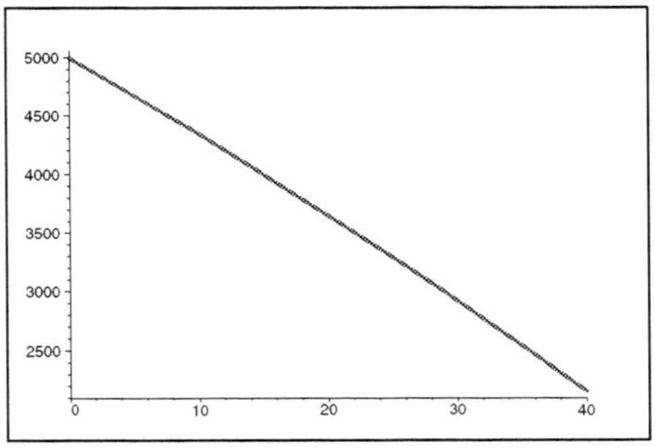

Figure 1. Size of population – $x(t)$.

In the figures you can see the difference in parameters in the case of cooperation (dark line) and in the case of the deviation on interval $[t_0, t_0 + 1]$ (light line). Fig. 1 shows population's dynamics. Fig. 2 shows territory sharing (s). You can see that s decreases from 0.5 to 0.3 at the interval $[t_0, t_0 + 1]$. Fig. 3 and 4 present the first and the second

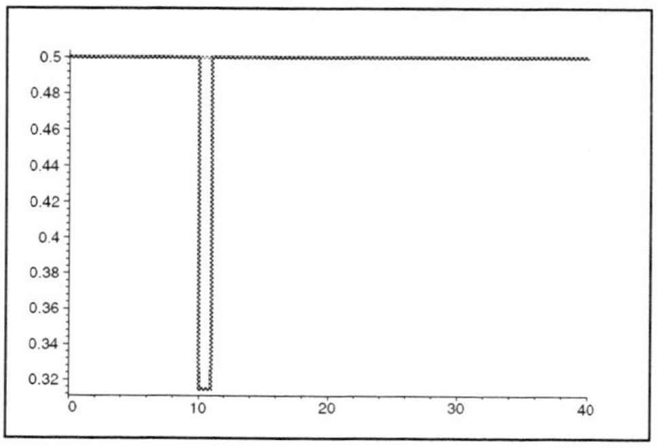

Figure 2. Territory sharing – $s(t)$.

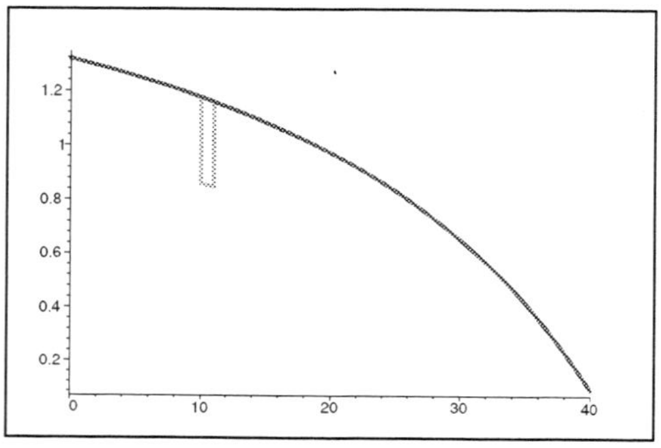

Figure 3. The first player's control – $E_1(t)$.

player's controls. We notice that the second player increases his fishing efforts and the first player decreases it at the time interval $[t_0, t_0 + 1]$. Fig. 5 and 6 present the first and the second player's catch, respectively. We notice that the first player's catch almost doesn't change, while the second player's catch decreases quickly at the time interval $[t_0, t_0 + 1]$.

The players' profits under cooperation are

$$J_1^d = 5233.346\,, \quad J_2^d = 5776.472\,,$$

and if the second player deviates the players' profits are

$$J_1^{otk} = 5233.346\,, \quad J_2^{otk} = 5749.256\,.$$

Time-consistent imputation distribution procedure.

Fig. 7 and 8 present time-consistent imputation distribution procedures for both players. So if players receive the gain according to this scheme they have no incentive to deviate from the cooperative agreement.

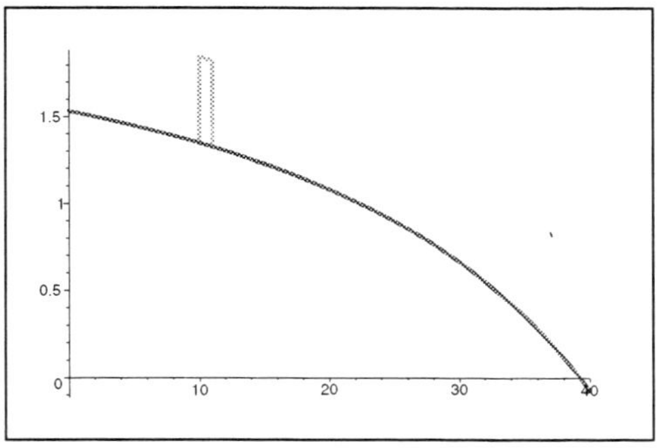

Figure 4. The second player's control – $E_2(t)$.

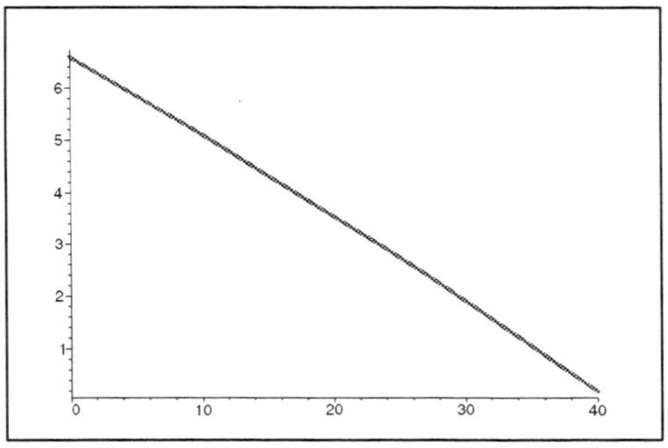

Figure 5. The first player's catch – $q_1 E_1(t)(1-s)x(t)$.

4. Discrete-Time Models

4.1. Bioresource Management Problem with Infinite Horizon

We consider here a discrete-time game model related with the bioresource management problem (fish catching). The center (referee) shares a reservoir between the competitors. The players (countries) which harvest the fish stock are the participants of this game.

Let us divide the water area into two parts: s and $1-s$, where two countries exploit the fish stock. The center (referee) shares the reservoir. The players (countries) which exploit the fish stock on their territory are the participants of this game.

The fish population grows according to the biological rule

$$x_{t+1} = (\varepsilon x_t)^{\alpha}, \quad x_0 = x, \tag{4.1}$$

where $x_t \geq 0$ – size of the population at a time t; $0 < \varepsilon < 1$ – natural death rate; $0 < \alpha < 1$ – natural birth rate.

We suppose that the utility function of country i is logarithmic:

$$I_1 = \ln((1-s)x_t u_t^1), \quad I_2 = \ln(s x_t u_t^2),$$

where u_t^1, $u_t^2 \geq 0$ – countries' fishing efforts at time t. We don't put the restriction $u_t^i \leq 1$, because the situation $u_t^i > 1$ is possible when country i enters the territory of the opponent.

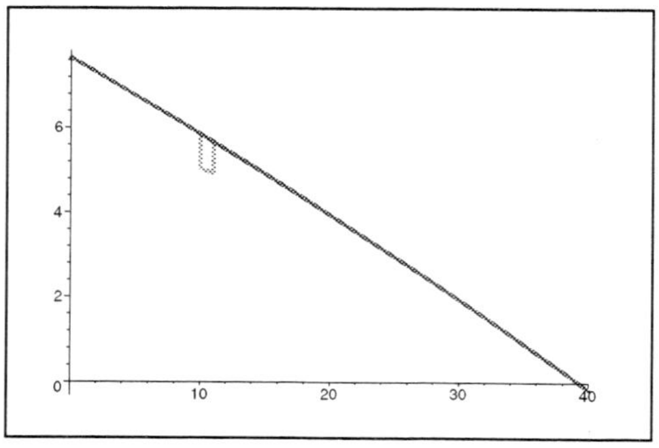

Figure 6. The second player's catch – $q_2 E_2(t) s x(t)$.

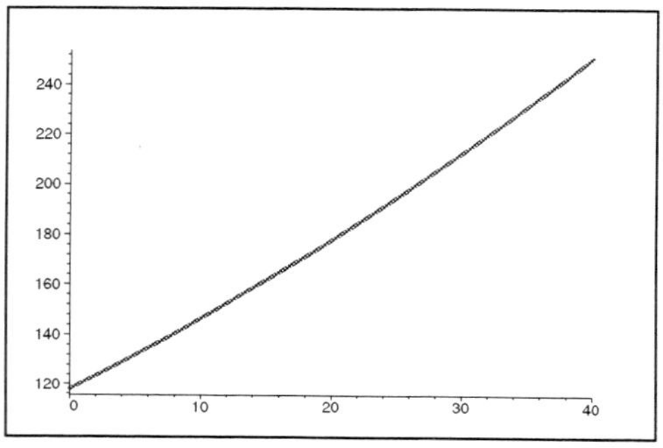

Figure 7. IDP for player 1 – $\beta_1(t)$.

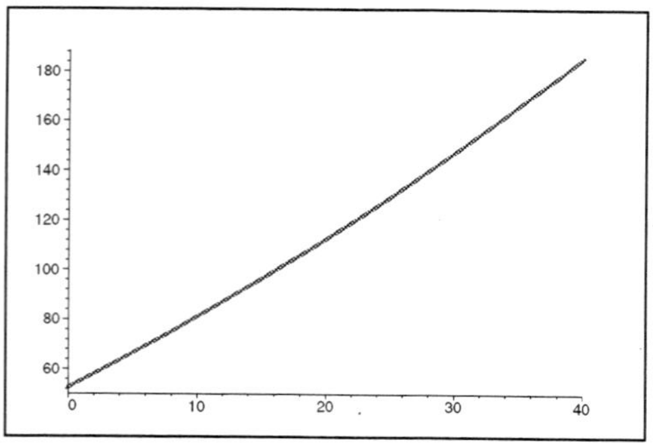

Figure 8. IDP for player 2 – $\beta_2(t)$.

We consider the problem of maximizing the sum of discounted utilities for two players:

$$J_1 = \sum_{t=0}^{n} \beta_1^t \ln((1-s)x_t u_t^1), \quad J_2 = \sum_{t=0}^{n} \beta_2^t \ln(s x_t u_t^2), \tag{4.2}$$

where $0 < \beta_i < 1$ – the discount factor for country i, $i = 1, 2$.

Also we consider the problem with infinite time horizon, here the players' net revenues are

$$\bar{J}_1 = \sum_{t=0}^{\infty} \beta_1^t \ln((1-s)x_t u_t^1)\,, \quad \bar{J}_2 = \sum_{t=0}^{\infty} \beta_2^t \ln(s x_t u_t^2)\,. \tag{4.3}$$

We obtain the Nash equilibrium and steady-state solution for the infinite-time problem (4.1), (4.3) in the next statement.

Theorem 4.1. *The Nash equilibrium for the n-step game (4.1), (4.2) is*

$$u_1^n = \frac{\varepsilon\alpha\beta_2 \sum\limits_{j=0}^{n-1} (\alpha\beta_2)^j}{(\sum\limits_{j=0}^{n} (\alpha\beta_1)^j \sum\limits_{j=0}^{n} (\alpha\beta_2)^j - 1)(1-s)}\,, \quad u_2^n = \frac{\varepsilon\alpha\beta_1 \sum\limits_{j=0}^{n-1} (\alpha\beta_1)^j}{(\sum\limits_{j=0}^{n} (\alpha\beta_1)^j \sum\limits_{j=0}^{n} (\alpha\beta_2)^j - 1)s}\,.$$

The Nash equilibrium for the infinite-time problem (4.1), (4.3) is

$$\bar{u}_1 = \frac{\varepsilon\beta_2(1-\alpha\beta_1)}{(\beta_1+\beta_2-\alpha\beta_1\beta_2)(1-s)}\,, \quad \bar{u}_2 = \frac{\varepsilon\beta_1(1-\alpha\beta_2)}{(\beta_1+\beta_2-\alpha\beta_1\beta_2)s}\,.$$

The dynamics of the fishery becomes

$$x_t = \left(\frac{\varepsilon\alpha\beta_1\beta_2}{\beta_1+\beta_2-\alpha\beta_1\beta_2}\right)^{\sum\limits_{j=1}^{t} \alpha^j} x_0^{\alpha^t}\,,$$

and the steady-state size of the population under Nash policies is

$$\bar{x} = \left(\frac{\varepsilon\alpha\beta_1\beta_2}{\beta_1+\beta_2-\alpha\beta_1\beta_2}\right)^{\frac{\alpha}{1-\alpha}}\,.$$

Proof. We use the approach (Levhari & Mirman, 1982). At first we consider the problem in context of the finite horizon. Assume that if there were no future period, the countries would get the remaining fish in the ratio $(1-s) : s$. Let the initial size of the population be x.

At first we consider one-step game. Suppose that the players fishing efforts are u_1 and u_2. These strategies must satisfy the inequality

$$u_1(1-s) + u_2 s < \varepsilon\,. \tag{4.4}$$

Then the first player's profit is

$$\begin{aligned} H_1^1(u_1, u_2) &= \ln((1-s)xu_1) + \beta_1 \ln((1-s)(\varepsilon x - (1-s)xu_1 - sxu_2)^{\alpha}) = \\ &= \ln((1-s)xu_1) + \alpha\beta_1 \ln(\varepsilon x - (1-s)xu_1 - sxu_2) + \beta_1 \ln(1-s)\,, \end{aligned}$$

and, similarly, the second player's profit is

$$H_2^1(u_1, u_2) = \ln(sxu_2) + \alpha\beta_2 \ln(\varepsilon x - (1-s)xu_1 - sxu_2) + \beta_2 \ln s\,.$$

Functions $H_1^1(u_1, u_2)$ and $H_2^1(u_1, u_2)$ are convex, so the Nash equilibrium exists and can be found from the system of equations $\partial H_1^1/\partial u_1 = 0$, $\partial H_2^1/\partial u_2 = 0$. So we find the Nash equilibrium for one-step game

$$u_1^1 = \frac{\varepsilon\alpha\beta_2}{(\alpha\beta_1+\alpha\beta_2+\alpha^2\beta_1\beta_2)(1-s)} = \frac{\varepsilon\alpha\beta_2}{((1+\alpha\beta_1)(1+\alpha\beta_2)-1)(1-s)},$$

$$u_2^1 = \frac{\varepsilon\alpha\beta_1}{(\alpha\beta_1+\alpha\beta_2+\alpha^2\beta_1\beta_2)s} = \frac{\varepsilon\alpha\beta_1}{((1+\alpha\beta_1)(1+\alpha\beta_2)-1)s}.$$

Now we verify that (4.4) is satisfied

$$u_1^1(1-s) + u_2^1 s = \frac{\varepsilon(\alpha\beta_1+\alpha\beta_2)}{\alpha\beta_1+\alpha\beta_2+\alpha^2\beta_1\beta_2} < \varepsilon\,.$$

The remaining stock of the fish is given by

$$\varepsilon x - (1-s)xu_1^1 - sxu_2^1 = \frac{\varepsilon\alpha^2\beta_1\beta_2}{(1+\alpha\beta_1)(1+\alpha\beta_2)-1}x\,.$$

Now we calculate the player 1's profit in this one-step game

$$\begin{aligned} H_1^1(u_1^1, u_2^1) &= \ln((1-s)xu_1^1) + \alpha\beta_1\ln(\varepsilon x - (1-s)xu_1^1 - sxu_2^1) + \beta_1\ln(1-s) = \\ &= (1+\alpha\beta_1)\ln x + A_1^1 + \beta_1\ln(1-s)\,, \end{aligned}$$

where A_1^1 is independent of x and is equal

$$A_1^1 = \ln\Big[\frac{(\varepsilon\alpha\beta_2)(\varepsilon\alpha^2\beta_1\beta_2)^{\alpha\beta_1}}{(\alpha\beta_1+\alpha\beta_2+\alpha^2\beta_1\beta_2)^{1+\alpha\beta_1}}\Big]\,.$$

And for player 2

$$\begin{aligned} H_2^1(u_1^1, u_2^1) &= \ln(sxu_2^1) + \alpha\beta_2\ln(\varepsilon x - (1-s)xu_1^1 - sxu_2^1) + \beta_2\ln s = \\ &= (1+\alpha\beta_2)\ln x + A_2^1 + \beta_2\ln s\,, \end{aligned}$$

where

$$A_2^1 = \ln\Big[\frac{(\varepsilon\alpha\beta_1)(\varepsilon\alpha^2\beta_1\beta_2)^{\alpha\beta_2}}{(\alpha\beta_1+\alpha\beta_2+\alpha^2\beta_1\beta_2)^{1+\alpha\beta_2}}\Big]\,.$$

Hence the objective function of player 1 for the two-step game is

$$H_1^2(u_1,u_2) = \ln((1-s)xu_1) + \alpha\beta_1(1+\alpha\beta_1)\ln(\varepsilon x-(1-s)xu_1-sxu_2) + \alpha\beta_1(A_1^1+\beta_1\ln(1-s))\,,$$

and for player 2

$$H_2^2(u_1,u_2) = \ln(sxu_2) + \alpha\beta_2(1+\alpha\beta_2)\ln(\varepsilon x-(1-s)xu_1-sxu_2) + \alpha\beta_2(A_2^1+\beta_2\ln s)\,.$$

Analogously, these functions are convex and we find the Nash equilibrium for the two-step game

$$u_1^2 = \frac{\varepsilon\alpha\beta_2(1+\alpha\beta_2)}{((1+\alpha\beta_1+\alpha^2\beta_1^2)(1+\alpha\beta_2+\alpha^2\beta_2^2)-1)(1-s)},$$

$$u_2^2 = \frac{\varepsilon\alpha\beta_1(1+\alpha\beta_1)}{((1+\alpha\beta_1+\alpha^2\beta_1^2)(1+\alpha\beta_2+\alpha^2\beta_2^2)-1)s}\,.$$

Also the restriction (4.4) is satisfied

$$u_1^2(1-s)+u_2^2 s = \frac{\varepsilon\alpha(\beta_1+\beta_2+\alpha\beta_1^2+\alpha\beta_2^2)}{(1+\alpha\beta_1+\alpha^2\beta_1^2)(1+\alpha\beta_2+\alpha^2\beta_2^2)-1} < \varepsilon\,.$$

Analogously we calculate the player 1's profit in the two-step game

$$H_1^2(u_1^2,u_2^2) = \ln((1-s)xu_1^2) + \alpha\beta_1(1+\alpha\beta_1)\ln(\varepsilon x-(1-s)xu_1^2 - sxu_2^2) +$$
$$+\alpha\beta_1(A_1^1+\beta_1\ln(1-s)) = (1+\alpha\beta_1+\alpha^2\beta_1^2)\ln x + A_1^2 + \alpha\beta_1(A_1^1+\beta_1\ln(1-s))\,,$$

where A_1^2 is equal

$$A_1^2 = \ln\frac{(\varepsilon(\alpha\beta_2+\alpha^2\beta_2^2)^{1+\alpha\beta_1+\alpha^2\beta_1^2}(\alpha\beta_1+\alpha^2\beta_1^2)^{\alpha\beta_1+\alpha^2\beta_1^2}}{((1+\alpha\beta_1+\alpha^2\beta_1^2)(1+\alpha\beta_2+\alpha^2\beta_2^2)-1)^{1+\alpha\beta_1+\alpha^2\beta_1^2}}\,.$$

And for player 2

$$H_2^1(u_1^1,u_2^1) = \ln(sxu_2^2) + \alpha\beta_2(1+\alpha\beta_2)\ln(\varepsilon x-(1-s)xu_1^2 - sxu_2^2) +$$
$$+\alpha\beta_2(A_2^1+\beta_2\ln s) = (1+\alpha\beta_2+\alpha^2\beta_2^2)\ln x + A_2^2 + \alpha\beta_2(A_2^1+\beta_2\ln s)\,,$$

where

$$A_2^2 = \ln\frac{(\varepsilon(\alpha\beta_1+\alpha^2\beta_1^2)^{1+\alpha\beta_2+\alpha^2\beta_2^2}(\alpha\beta_2+\alpha^2\beta_2^2)^{\alpha\beta_2+\alpha^2\beta_2^2}}{((1+\alpha\beta_1+\alpha^2\beta_1^2)(1+\alpha\beta_2+\alpha^2\beta_2^2)-1)^{1+\alpha\beta_2+\alpha^2\beta_2^2}}\,.$$

The process can be repeated for the n-step game and we find the Nash equilibrium in the form, that is given in the statement. And the restriction (4.4) is satisfied because

$$\sum_{j=0}^{n}(\alpha\beta_1)^j\sum_{j=0}^{n}(\alpha\beta_2)^j - 1 = \sum_{j=1}^{n}(\alpha\beta_1)^j + \sum_{j=1}^{n}(\alpha\beta_2)^j + \sum_{j=1}^{n}(\alpha\beta_1)^j\sum_{j=1}^{n}(\alpha\beta_2)^j >$$
$$> \sum_{j=1}^{n}(\alpha\beta_1)^j + \sum_{j=1}^{n}(\alpha\beta_2)^j = \alpha\beta_1\sum_{j=0}^{n-1}(\alpha\beta_1)^j + \alpha\beta_2\sum_{j=0}^{n-1}(\alpha\beta_2)^j\,.$$

Then we tend the horizon to infinity and receive the Nash equilibrium for the infinite-time problem (4.1), (4.3) (Nowak, 2006). So we receive

$$\bar{u}_1 = \frac{\varepsilon\beta_2(1-\alpha\beta_1)}{(\beta_1+\beta_2-\alpha\beta_1\beta_2)(1-s)}\,,$$

$$\bar{u}_2 = \frac{\varepsilon\beta_1(1-\alpha\beta_2)}{(\beta_1+\beta_2-\alpha\beta_1\beta_2)s}\,,$$

$$\varepsilon x - \bar{u}_1(1-s)x - \bar{u}_2 s x = \frac{\varepsilon\alpha\beta_1\beta_2}{\beta_1+\beta_2-\alpha\beta_1\beta_2}x\,,$$

and the inequality (4.4) is satisfied

$$\bar{u}_1(1-s)+\bar{u}_2 s = \frac{\varepsilon(\beta_1+\beta_2-2\alpha\beta_1\beta_2)}{\beta_1+\beta_2-\alpha\beta_1\beta_2} < \varepsilon\,.$$

Now we find the steady-state solution. Under Nash equilibrium strategies the dynamics of the fish population is

$$x_{t+1} = (\varepsilon x_t - \bar{u}_1(1-s)x_t - \bar{u}_2 s x_t)^\alpha = \left(\frac{\varepsilon\alpha\beta_1\beta_2}{\beta_1+\beta_2-\alpha\beta_1\beta_2}\right)^{\sum_{j=1}^{t}\alpha^j} x_0^{\alpha^t}.$$

And steady-state is (as $n \to \infty$)

$$\bar{x} = \left(\frac{\varepsilon\alpha\beta_1\beta_2}{\beta_1+\beta_2-\alpha\beta_1\beta_2}\right)^{\frac{\alpha}{1-\alpha}}.$$

Cooperative equilibrium. We find the cooperative equilibrium using the same approach of transfer from finite to infinite resource management problem. Here the players wish to maximize the discounted sum of their total utilities on finite and infinite time horizon:

$$\max_{u_t^1, u_t^2} \sum_{t=0}^{n} \rho^t (\mu_1 \ln((1-s)x_t u_t^1) + \mu_2 \ln(s x_t u_t^2)), \tag{4.5}$$

$$\max_{u_t^1, u_t^2} \sum_{t=0}^{\infty} \rho^t (\mu_1 \ln((1-s)x_t u_t^1) + \mu_2 \ln(s x_t u_t^2)), \tag{4.6}$$

where $0 < \rho < 1$ – the common discount factor, $0 < \mu_1, \mu_2 < 1$ – weighting coefficients ($\mu_1 + \mu_2 = 1$).

Theorem 4.2. *The cooperative equilibrium for n-step game (4.1), (4.5) is*

$$u_1^{dn} = \frac{\varepsilon\mu_1}{\sum_{j=0}^{n}(\alpha\rho)^j(1-s)}, \qquad u_2^{dn} = \frac{\varepsilon\mu_2}{\sum_{j=0}^{n}(\alpha\rho)^j s}.$$

The cooperative equilibrium for the problem (4.1), (4.6) is

$$u_1^d = \frac{\varepsilon\mu_1(1-\alpha\rho)}{1-s}, \qquad u_2^d = \frac{\varepsilon\mu_2(1-\alpha\rho)}{s}.$$

The dynamics of the fishery becomes

$$x_t^d = (\varepsilon\alpha\rho)^{\sum_{j=1}^{t}\alpha^j} x_0^{\alpha^t},$$

and the steady-state size of the population under cooperative policies is

$$\bar{x}^d = (\varepsilon\alpha\rho)^{\frac{\alpha}{1-\alpha}}.$$

Proof. The proof of this statement is analogous to Theorem 4.1, so we don't give it here. We remark that the restriction (4.4) is satisfied for finite:

$$u_1^{dn}(1-s) + u_2^{dn} s = \frac{\varepsilon}{\sum_{j=0}^{n}(\alpha\rho)^j} < \varepsilon,$$

and for infinite horizon:

$$u_1^d(1-s)+u_2^d s=\varepsilon(1-\alpha\rho)<\varepsilon\,.$$

Incentive equilibrium. Here it appears the problem of maintenance the agreement which was achieved at the beginning of the game and punishing the player who deviates. In this section we consider the cooperative incentive equilibrium as the method of keeping cooperation, moreover the center (referee) punishes the players for deviation.

Let's denote s^d – water area sharing under cooperation.

We find the cooperative incentive equilibrium for n-step game and for infinite-time game.

Theorem 4.3. *The cooperative incentive equilibrium of the problem (4.1), (4.5) is*

$$\gamma_1^n(u_2)=\frac{\varepsilon\mu_1}{\sum\limits_{j=0}^{n}(\alpha\rho)^j(1-s_2^{*n})}\,,\qquad \gamma_2^n(u_1)=\frac{\varepsilon\mu_2}{\sum\limits_{j=0}^{n}(\alpha\rho)^j s_1^{*n}}\,,$$

where

$$s_2^{*n}=s^d-\frac{s^d}{u_2^{dn}}(u_2-u_2^{dn})\,,\quad s_1^{*n}=s^d+\frac{1-s^d}{u_1^{dn}}(u_1-u_1^{dn})\,.$$

The cooperative incentive equilibrium of the problem (4.1), (4.6) is

$$\gamma_1(u_2)=\frac{\varepsilon\mu_1(1-\alpha\rho)}{1-s_2^*}\,,\qquad \gamma_2(u_1)=\frac{\varepsilon\mu_2(1-\alpha\rho)}{s_1^*}\,,$$

where

$$s_2^*=s^d-\frac{s^d}{u_2^d}(u_2-u_2^d)\,,\quad s_1^*=s^d+\frac{1-s^d}{u_1^d}(u_1-u_1^d)\,.$$

Proof. We assume that the center punishes the players for a deviation from the equilibrium point, but not themselves, as it was in (Ehtamo & Hamalainen, 1993). Assume than, if the first player deviates the center increases s, but if the second player deviates – decreases s.

Let consider the second player's deviation for the n-step game

$$u_2^n=u_2^{dn}+\Delta\,.$$

We find the center's strategy in the following form

$$s^{*n}=s^d-\eta^n(u_2^n-u_2^{dn})\,.$$

Then the first player's punishment strategy is

$$\gamma_1^n(u_2^n)=\frac{\varepsilon\mu_1}{\sum\limits_{j=0}^{n}(\alpha\rho)^j(1-s^{*n})}\,.$$

Next we should determine the coefficient η^n. For that we solve the problem for player 2 assuming that player 1 uses the punishment strategy, so

$$\sum_{t=0}^{n} \rho^t \ln(s^{*t} x u_2^t),$$

where

$$s^{*t} = s^d - \eta^t (u_2^t - u_2^{dt}),$$

and dynamics of the fishery becomes

$$x_{t+1} = (\varepsilon x_t - \gamma_1^t(u_2^t)(1 - s^{*t}) x_t - s^{*t} x_t u_2^t)^\alpha .$$

For η^n to be the incentive equilibrium the solution of this optimization problem must be achieved on u_2^{dn}.

For solving that problem we use the same approach as above. So at first we consider one-step game

$$u_2 = u_2^{d1} + \Delta, \quad s^{*1} = s^d - \eta^1 (u_2 - u_2^{d1}),$$

$$\gamma_1^1(u_2) = \frac{\varepsilon \mu_1}{(1 + \alpha\rho)(1 - s^{*1})}.$$

The objective function for player 2 becomes

$$\max_{u_2 \geq 0} \{\ln(s^{*1} x u_2) + \alpha\rho \ln(\varepsilon x - \gamma_1^1(u_2)(1 - s^{*1}) x - s^{*1} x u_2)\}$$

and the solution of this problem u_2 must be equal

$$u_2^{d1} = \frac{\varepsilon \mu_2}{(1 + \alpha\rho) s^d},$$

from there we find that

$$\eta^1 = \frac{s^d}{u_2^{d1}}.$$

Continuing this process for n-step game we receive

$$\eta^n = \frac{s^d}{u_2^{dn}}.$$

Here inequality (4.4) also satisfies, because

$$\gamma_1^t(u_2^t)(1 - s^{*t}) + s^{*t} x_t = \frac{\varepsilon \mu_1}{\sum\limits_{j=0}^{t} (\alpha\rho)^j} + u_2^t s^d \left(2 - \frac{u_2^t}{u_2^{dt}}\right) = \frac{\varepsilon \mu_1}{\sum\limits_{j=0}^{t} (\alpha\rho)^j} + 2 u_2^t s^d -$$

$$- \frac{(u_2^t)^2 (s^d)^2 \sum\limits_{j=0}^{t} (\alpha\rho)^j}{\varepsilon \mu_2} = \frac{\varepsilon}{\sum\limits_{j=0}^{t} (\alpha\rho)^j} - \frac{(u_2^t s^d \sum\limits_{j=0}^{t} (\alpha\rho)^j - \varepsilon \mu_2)^2}{\varepsilon \mu_2 \sum\limits_{j=0}^{t} (\alpha\rho)^j} < \varepsilon .$$

Analogously for player 1 we find the center's strategy in the following form

$$s^{*n} = s^d + \theta^n (u_1^n - u_1^{dn}) .$$

Then the second player's punishment strategy is

$$\gamma_2^n(u_1^n) = \frac{\varepsilon \mu_2}{\sum\limits_{j=0}^{n} (\alpha\rho)^j s^{*n}} .$$

To determine the coefficient θ^n we solve the problem for player 1 assuming that player 2 uses the punishment strategy, so

$$\sum_{t=0}^{n} \rho^t \ln((1 - s^{*t})x_t u_1^t) ,$$

where

$$s^{*t} = s^d + \theta^t (u_1^t - u_1^{dt}) ,$$

and dynamics of the fishery becomes

$$x_{t+1} = (\varepsilon x_t - u_1^t(1 - s^{*t})x_t - \gamma_2^t(u_1^t)s^{*t}x_t)^\alpha .$$

So we receive that

$$\theta^n = \frac{1 - s^d}{u_1^{dn}} .$$

The incentive equilibrium of problem (4.1), (4.6) we get letting the horizon tend to infinity.

Corollary 4. 1. *The same cooperative incentive strategies are in the case when there are more than one deviation.*

Proof. Consider two-step game with two deviation. Then for one-step game

$$u_2 = u_2^{d1} + \Delta_1 , \quad s^{*1} = s^d - \eta^1(u_2 - u_2^{d1}) ,$$

$$\gamma_1^1(u_2) = \frac{\varepsilon \mu_1}{(1 + \alpha\rho)(1 - s^{*1})}$$

and the objective function for player 2 is

$$\max_{u_2 \geq 0}\{\ln(s^* 1 x u_2) + \alpha\rho \ln(\varepsilon x - \gamma_1^1(u_2)(1 - s^{*1})x - s^{*1}xu_2)\} .$$

Now we consider two-step game under assumption that in the following one-step game there is a deviation $u_2 = u_2^{d1} + \Delta_1$.

Then the objective function for the two-step game is

$$\begin{aligned}
&\max_{u_2 \geq 0}\{\ln(s^{*2}xu_2) + \rho(\text{profit in one-step game as } x = x_2 = (\varepsilon x - \gamma_1^2(u_2)(1 - s^{*2})x - s^{*2}xu_2)^\alpha\} = \\
&= \max_{u_2 \geq 0}\{\ln(s^{*2}xu_2) + \rho(\ln(s^{*1}x_2u_2^1) + \alpha\rho \ln(\varepsilon x_2 - \gamma_1^1(u_2)(1 - s^{*1})x_2 - s^{*1}x_2u_2^1))\} = \\
&= \max_{u_2 \geq 0}\{\ln(s^{*2}xu_2) + \rho(\ln(s^{*1}u_2^1) + \ln x_2 + \alpha\rho \ln(\varepsilon - \gamma_1^1(u_2)(1 - s^{*1}) - s^{*1}u_2^1) + \\
&+ \alpha\rho \ln x_2)\} = \max_{u_2 \geq 0}\{\ln(s^{*2}xu_2) + \alpha\rho(1 + \alpha\rho) \ln(\varepsilon x - \gamma_1^2(u_2)(1 - s^{*2})x - s^{*2}xu_2)\} + \\
&+ \rho \max_{u_2^1 \geq 0}\{\ln(s^{*1}u_2^1) + \alpha\rho \ln(\varepsilon - \gamma_1^1(u_2)(1 - s^{*1}) - s^{*1}u_2^1)\} .
\end{aligned}$$

It's not difficult to verify that the maximum is achieved at the points u_2^{d1}, u_2^{d2} when $\eta_2^1 = \frac{s^d}{u_2^{d1}}$, $\eta_2^2 = \frac{s^d}{u_2^{d2}}$.

We investigate optimal strategies in the case of of short-time second player's deviation on step k and his returning to cooperation after.

Corollary 4.2. *The steady-state size of the population under deviation is equal to cooperative case when the number of steps tends to infinity:*

$$x_n^{dev} \to (\varepsilon\alpha\rho)^{\frac{\alpha}{1-\alpha}}.$$

Proof. Let player 2 deviates on step k

$$u_2^k = u_2^{dk} + \Delta_k$$

and after he continues to use cooperative strategy u_2^{dn}.

So the player 1 punishes the player 2 on step k according to Theorem 4.3

$$\gamma_1^k(u_2^k) = \frac{\varepsilon\mu_1}{\sum\limits_{j=0}^{k}(\alpha\rho)^j(1 - s^d + \frac{s^d}{u_2^{dk}})(u_2^k - u_2^{dk})}.$$

Now we find the population dynamic with deviation (x_t^{dev}) and without it (x_t^d).
It is not difficult to receive that

$$x_k^d = x_k^{dev} = x^{\alpha^k}\varepsilon^{\sum\limits_{j=1}^{k}\alpha^j}\prod_{p=1}^{k}\Big(\frac{\sum\limits_{j=1}^{p}(\alpha\rho)^j}{\sum\limits_{j=0}^{p}(\alpha\rho)^j}\Big)^{\alpha^{n-p+1}}.$$

On step $k+1$

$$x_{k+1}^d = (x_k^d)^\alpha\Big(\varepsilon\frac{\sum\limits_{j=1}^{k}(\alpha\rho)^j}{\sum\limits_{j=0}^{k}(\alpha\rho)^j}\Big)^\alpha, \quad x_{k+1}^{dev} = (x_k^d)^\alpha D^\alpha,$$

where

$$D = \varepsilon\frac{\sum\limits_{j=1}^{k}(\alpha\rho)^j}{\sum\limits_{j=0}^{k}(\alpha\rho)^j} + \frac{(s^d\Delta_k)^2\sum\limits_{j=0}^{k}(\alpha\rho)^j}{\varepsilon\mu_2}.$$

Continuing process from step $k+2$ we receive

$$x_n^d = x^{\alpha^n}\varepsilon^{\sum\limits_{j=1}^{n}\alpha^j}\prod_{p=1}^{n}\Big(\frac{\sum\limits_{j=1}^{p}(\alpha\rho)^j}{\sum\limits_{j=0}^{p}(\alpha\rho)^j}\Big)^{\alpha^{n-p+1}} = x^{\alpha^n}(\varepsilon\alpha\rho)^{\sum\limits_{j=1}^{n}\alpha^j}\prod_{p=1}^{n}\Big(\frac{1-(\alpha\rho)^p}{1-(\alpha\rho)^{p+1}}\Big)^{\alpha^{n-p+1}},$$

$$x_n^{dev} = x^{\alpha^n} \varepsilon^{\sum\limits_{j=1}^{n}\alpha^j} \prod_{p=1}^{n}\left(\frac{\sum\limits_{j=1}^{p}(\alpha\rho)^j}{\sum\limits_{j=0}^{p}(\alpha\rho)^j}\right)^{\alpha^{n-p+1}} \left(\frac{D}{\varepsilon}\frac{\sum\limits_{j=0}^{k}(\alpha\rho)^j}{\sum\limits_{j=1}^{k}(\alpha\rho)^j}\right)^{\alpha^{n-k}} =$$

$$= x^{\alpha^n}(\varepsilon\alpha\rho)^{\sum\limits_{j=1}^{n}\alpha^j} \prod_{p=1}^{n}\left(\frac{1-(\alpha\rho)^p}{1-(\alpha\rho)^{p+1}}\right)^{\alpha^{n-p+1}} \left(1+\frac{(s^d\Delta_k\sum\limits_{j=0}^{k}(\alpha\rho)^j)^2}{\varepsilon^2\mu_2\sum\limits_{j=1}^{k}(\alpha\rho)^j}\right)^{\alpha^{n-k}}. \quad (4.7)$$

We need to show that as n tends to infinity

$$x_n^d \to (\varepsilon\alpha\rho)^{\frac{\alpha}{1-\alpha}}, \quad x_n^{dev} \to (\varepsilon\alpha\rho)^{\frac{\alpha}{1-\alpha}}.$$

For that we show that as $n \to \infty$

$$P = \prod_{p=1}^{n}\left(\frac{1-(\alpha\rho)^p}{1-(\alpha\rho)^{p+1}}\right)^{\alpha^{n-p+1}} \to 1.$$

Let's consider

$$\ln P = \sum_{p=1}^{n}\alpha^{n-p+1}[\ln(1-(\alpha\rho)^p) - \ln(1-(\alpha\rho)^{p+1})] = -\sum_{p=1}^{n}\alpha^{n-p+1}\ln(\tfrac{1-(\alpha\rho)^{p+1})}{1-(\alpha\rho)^{p+1}}) =$$
$$= -\sum_{p=1}^{n}\alpha^{n-p+1}\ln(1+\tfrac{(\alpha\rho)^p(1-\alpha\rho)}{1-(\alpha\rho)^p})..$$

Since $\ln(1+x) \le x$ then

$$\ln P \le -\sum_{p=1}^{n}\alpha^{n-p+1}\frac{(\alpha\rho)^p(1-\alpha\rho)}{1-(\alpha\rho)^p}.$$

We find the limit as

$$\lim_{n\to\infty}\sum_{p=1}^{n}\alpha^{n-p+1}\tfrac{(\alpha\rho)^p(1-\alpha\rho)}{1-(\alpha\rho)^p} = \lim_{n\to\infty}\alpha^{n+1}(1-\alpha\rho)\sum_{p=1}^{n}\tfrac{\rho^p}{1-(\alpha\rho)^p} =$$
$$= (1-\alpha\rho)\lim_{n\to\infty}\tfrac{\rho(\alpha\rho)^{n+1}}{(n+1)\ln(1/\alpha)} = 0.$$

It yields

$$\lim_{n\to\infty}\ln P \le 0.$$

Note also that as $P \ge 1$, then $\ln P \ge 0$.

Finally we get

$$\lim_{n\to\infty} P = 1.$$

Corollary 4.3. *The conditions for incentive equilibrium are satisfied*

$$J_1^{dev} \ge J_1^d, \quad J_2^{dev} \le J_2^d,$$

where J_i^d – the profit of the player i when both players use cooperative strategies, J_i^{dev} – the profit of the player i when the second player deviates and the center punishes him $(i = 1, 2)$.

Proof. Let again player 2 deviates on step k

$$u_2^k = u_2^{dk} + \Delta_k$$

and after he continues to use cooperative strategy u_2^{dn}.

So the player 1 punishes the player 2 on step k according to Theorem 4.3

$$\gamma_1^k(u_2^k) = \frac{\varepsilon\mu_1}{\sum\limits_{j=0}^{k}(\alpha\rho)^j(1-s^d+\frac{s^d}{u_2^{dk}})(u_2^k-u_2^{dk})}.$$

Let's find first player's profit under second player's deviation

$$J_1^{dev} = \sum_{t=0}^{k-1}\rho^t\ln((1-s^d)x_t^d u_1^{dt}) + \rho^k\ln\Big(x_k^d\frac{\varepsilon\mu_1}{\sum\limits_{j=0}^{k}(\alpha\rho)^j}\Big) + \sum_{t=k+1}^{n}\rho^t\ln((1-s^d)x_t^{dev}u_1^{dt}).$$

Therefore

$$J_1^{dev} - J_1^d = \sum_{t=k+1}^{n}\rho^t\ln((1-s^d)x_t^{dev}u_1^{dt}) - \sum_{t=k+1}^{n}\rho^t\ln((1-s^d)x_t^d u_1^{dt}) = \sum_{t=k+1}^{n}\rho^t\ln\frac{x_t^{dev}}{x_t^d}.$$

Substituting (4.7) we get

$$J_1^{dev} - J_1^d =$$

$$= \sum_{t=k+1}^{n}\rho^t\alpha^{t-k}\ln\Big(1+\frac{(s^d\Delta_k\sum\limits_{j=0}^{k}(\alpha\rho)^j)^2}{\varepsilon^2\mu_2\sum\limits_{j=1}^{k}(\alpha\rho)^j}\Big) = \rho^k\sum_{j=1}^{n-k}(\alpha\rho)^j\ln\Big(1+\frac{(s^d\Delta_k\sum\limits_{j=0}^{k}(\alpha\rho)^j)^2}{\varepsilon^2\mu_2\sum\limits_{j=1}^{k}(\alpha\rho)^j}\Big).$$

From where we conclude that

$$J_1^{dev} \geq J_1^d.$$

For player 2

$$J_2^{dev} = \sum_{t=0}^{k-1}\rho^t\ln(s^d x_t^d u_2^{dt}) + \rho^k\ln(s^d x_k^d u_2^{dk} - \frac{s^d}{u_2^{dk}}x_k^d\Delta^2) + \sum_{t=k+1}^{n}\rho^t\ln(s^d x_t^{dev}u_2^{dt}).$$

$$\begin{aligned} J_2^{dev} - J_2^d &= \rho^k\ln(1-\frac{\Delta_k^2}{(u_2^{dk})^2}) + \sum_{t=k+1}^{n}\rho^t\ln\frac{x_t^{dev}}{x_t^d} = \\ &= \rho^k\ln\Big[(1-\frac{(s^d\Delta_k\sum\limits_{j=0}^{k}(\alpha\rho)^j)^2}{\varepsilon^2\mu_2^2})(1+\frac{(s^d\Delta_k\sum\limits_{j=0}^{k}(\alpha\rho)^j)^2}{\varepsilon^2\mu_2\sum\limits_{j=1}^{k}(\alpha\rho)^j})^{\sum\limits_{j=1}^{n-k}(\alpha\rho)^j}\Big]. \end{aligned}$$

Now we show that if $n \geq 2$

$$J_2^{dev} \leq J_2^d.$$

Let's denote $a = \frac{(s^d \Delta_k \sum_{j=0}^{k} (\alpha\rho)^j)^2}{\varepsilon^2 \mu_2}$. Since $\ln(1+x) \le x$ then

$$J_2^{dev} - J_2^d = \rho^k [\ln(1 - \tfrac{a}{\mu_2}) + \sum_{j=1}^{n-k} (\alpha\rho)^j \ln(1 + \frac{a}{\sum_{j=1}^{k} (\alpha\rho)^j})] \le$$

$$\le \rho^k \frac{a(\sum_{j=1}^{n-k} (\alpha\rho)^j \mu_2 - \sum_{j=1}^{k} (\alpha\rho)^j)}{\mu_2 \sum_{j=1}^{k} (\alpha\rho)^j} = -a\rho^k \frac{1-(\alpha\rho)^k - \mu_2 + \mu_2(\alpha\rho)^{n-k}}{\mu_2(1-(\alpha\rho)^k)}.$$

The right-hand side is nonpositive. To show it we consider the function

$$D(k) = 1 - (\alpha\rho)^k - \mu_2 + \mu_2(\alpha\rho)^{n-k}.$$

Note that $D(k)$ increases on k, because $\frac{\partial D}{\partial k} > 0$.
So it is sufficient to verify that $D(1) > 0$. It is equivalent to

$$\mu_2 < \frac{1 - \alpha\rho}{1 - (\alpha\rho)^{n-1}},$$

that holds when $n \ge 2$. From where we conclude that

$$J_2^{dev} \le J_2^d.$$

Corollary 4.4. *The player who deviates losses less when the number of steps grows*

$$D^{n+1} < D^n,$$

where $D^n = J_2^{dn} - J_2^{dev\, n}$.

Proof. We consider the value of player 2 loss in comparison with cooperative case for the n-step game and for the $n+1$-step game.

$$\Delta^n = J_2^{dev} - J_2^d = \rho^k \ln\Big[(1 - \frac{(s^d \Delta_k \sum_{j=0}^{k} (\alpha\rho)^j)^2}{\varepsilon^2 \mu_2^2})(1 + \frac{(s^d \Delta_k \sum_{j=0}^{k} (\alpha\rho)^j)^2}{\varepsilon^2 \mu_2 \sum_{j=1}^{k} (\alpha\rho)^j})^{\sum_{j=1}^{n-k} (\alpha\rho)^j}\Big].$$

Then

$$\Delta^{n+1} - \Delta^n = \rho^k \ln \frac{\Big(1 - \frac{(s^d \Delta_k \sum_{j=0}^{k} (\alpha\rho)^j)^2}{\varepsilon^2 \mu_2^2}\Big)\Big(1 + \frac{(s^d \Delta_k \sum_{j=0}^{k} (\alpha\rho)^j)^2}{\varepsilon^2 \mu_2 \sum_{j=1}^{k} (\alpha\rho)^j}\Big)^{\sum_{j=1}^{n+1-k} (\alpha\rho)^j}}{\Big(1 - \frac{(s^d \Delta_k \sum_{j=0}^{k} (\alpha\rho)^j)^2}{\varepsilon^2 \mu_2^2}\Big)\Big(1 + \frac{(s^d \Delta_k \sum_{j=0}^{k} (\alpha\rho)^j)^2}{\varepsilon^2 \mu_2 \sum_{j=1}^{k} (\alpha\rho)^j}\Big)^{\sum_{j=1}^{n-k} (\alpha\rho)^j}} =$$

$$= \rho^k (\alpha\rho)^{n+1-k} \ln\Big(1 + \frac{(s^d \Delta_k \sum_{j=0}^{k} (\alpha\rho)^j)^2}{\varepsilon^2 \mu_2 \sum_{j=1}^{k} (\alpha\rho)^j}\Big) > 0.$$

Conclude that

$$\Delta^{n+1} > \Delta^n,$$

but $\Delta^{n+1} < 0$ and $\Delta^n < 0$, so player 2 losses less in $n+1$-step game.

4.2. Discrete-Time Model with Migration

We consider here a discrete-time game model related to a bioresource management problem (fish catching). We divide a fishery into regions, which are exploited by single players. The center (referee) shares a reservoir into regions, where the players (countries) harvest the fish stock.

We assume that there are migratory exchanges between the regions. So the stock in the one region (where the player 1 exploits the fish) depends not only on the previous stock and catch in the region, but also on the stock and catch in other region (where the player 2 exploits the fish).

In this section, we also consider the case when the center is a player and forms a coalition with the players (countries). We find the Shapley value and time-consistent imputation distribution procedure for this model. We introduce a new condition which offers an incentive to players to cooperate and we call it the each step rational behavior condition.

Let us divide the water area into two parts: s and $1-s$, where two countries exploit the fish stock. The center (referee) shares the reservoir. Each player (country) can exploit the fish stock in only one of the two regions.

The fish populations grow according to the biological rule (Mazalov & Rettieva, 2010)

$$\begin{cases} x_{t+1} = x_t^{\alpha_1}\left(\frac{y_t}{x_t}\right)^{\beta_1 s}, & x_0 = x\,, \\ y_{t+1} = y_t^{\alpha_2}\left(\frac{x_t}{y_t}\right)^{\beta_2(1-s)}, & y_0 = y\,, \end{cases} \tag{4.8}$$

where $x_t \geq 0$ – size of the population in the first region at a time t, $y_t \geq 0$ – size of the population in the second region at a time t, $0 < \alpha_i < 1$ – natural birth rate, $0 < \beta_i < 1$ – coefficients of migration between the regions ($i = 1, 2$).

Here the α_i represent the direct effect of the stock on the next period's stock on this territory. The β_i represent the effect of migration between two parts of the reservoir. We suppose that $\alpha_i > \beta_i$, $i = 1, 2$, which corresponds to the case when internal natural birth rate is more powerful for the population growth than the migratory exchange.

Player 1 exploits the stock x_t and player 2 harvests the stock y_t.

You can see that in our model the intensity of migration also depends on the share s. It seems to be natural because the habitat size decreases as s decreases and fish needs to migrate to another region.

We suppose that the utility function of country i is logarithmic. Then the players' net revenues over infinite time horizon are:

$$J_1 = \sum_{t=0}^{\infty} \delta_1^t \ln(u_{1t})\,, \quad J_2 = \sum_{t=0}^{\infty} \delta_2^t \ln(u_{2t})\,, \tag{4.9}$$

where $0 \leq u_{1t} \leq x_t$, $0 \leq u_{2t} \leq y_t$ – countries' catch at a time t, $0 < \delta_i < 1$ – the discount factor for country i ($i = 1, 2$).

Each player is interested in maximizing the sum of his discounted utility. And the dynamics become

$$\begin{cases} x_{t+1} = (x_t - u_{1t})^{\alpha_1 - \beta_1 s}(y_t - u_{2t})^{\beta_1 s}\,, \\ y_{t+1} = (y_t - u_{2t})^{\alpha_2 - \beta_2(1-s)}(x_t - u_{1t})^{\beta_2(1-s)}\,. \end{cases} \tag{4.10}$$

Noncooperative case. We begin by analyzing the noncooperative case and find the Nash equilibrium. Also we determine the steady-state size of the populations.

In the next statement we find optimal players' controls (catch) in the noncooperative case.

Theorem 4.4. *The Nash equilibrium of the problem (4.9)-(4.10) is*

$$u_1 = \gamma_1 x\,, \quad u_2 = \gamma_2 y\,, \tag{4.11}$$

where

$$\gamma_1 = \frac{\Delta_1}{1-\delta_1\alpha_2+\delta_1\beta_2(1-s)}\,, \quad \gamma_2 = \frac{\Delta_2}{1-\delta_2\alpha_1+\delta_2\beta_1 s}\,,$$

$$\Delta_i = (1-\delta_i\alpha_2+\delta_i\beta_2(1-s))(1-\delta_i\alpha_1+\delta_i\beta_1 s) - \delta_i^2 s(1-s)\beta_1\beta_2\,,\ i=1,2.$$

Proof. Let $V_1(x,y)$ be the value function for player 1, and $V_2(x,y)$ – for player 2. Using the Bellman principle, these functions must satisfy

$$V_1(x,y) = \max_{0\le u_1\le x}\{\ln u_1 + \delta_1 V_1((x-u_1)^{\alpha_1}\left(\frac{y-u_2}{x-u_1}\right)^{\beta_1 s}, (y-u_2)^{\alpha_2}\left(\frac{x-u_1}{y-u_2}\right)^{\beta_2(1-s)})\}\,, \tag{4.12}$$

$$V_2(x,y) = \max_{0\le u_2\le y}\{\ln u_2 + \delta_2 V_2((x-u_1)^{\alpha_1}\left(\frac{y-u_2}{x-u_1}\right)^{\beta_1 s}, (y-u_2)^{\alpha_2}\left(\frac{x-u_1}{y-u_2}\right)^{\beta_2(1-s)})\}\,. \tag{4.13}$$

Let us assume that the value function has the following form:

$$V_i(x,y) = A_i\ln x + B_i\ln y + C_i\,, \quad i=1,2,$$

where A_i, B_i and C_i are constants that depend on the parameters of the model.

Then for player 1 from (4.12) we receive

$$\begin{aligned} A_1\ln x + B_1\ln y + C_1 &= \ln u_1 + \delta_1 A_1[(\alpha_1-\beta_1 s)\ln(x-u_1) + \beta_1 s\ln(y-u_2)] + \\ &+ \delta_1 B_1[(\alpha_2-\beta_2(1-s))\ln(y-u_2) + \beta_2(1-s)\ln(x-u_1)] + \delta_1 C_1\,. \end{aligned} \tag{4.14}$$

Suppose that $u_1 = \gamma_1 x$ and $u_2 = \gamma_2 y$. Then we receive the system of equations for parameters:

$$\left\{\begin{array}{rcl} A_1 &=& 1+\delta_1 A_1(\alpha_1-\beta_1 s)+\delta_1 B_1\beta_2(1-s)\,,\\ B_1 &=& \delta_1 A_1\beta_1 s + \delta_1 B_1(\alpha_2-\beta_2(1-s))\,,\\ C_1 &=& \ln\gamma_1 + \delta_1 A_1[(\alpha_1-\beta_1 s)\ln(1-\gamma_1)+\beta_1 s\ln(1-\gamma_2)]+\\ &+& \delta_1 B_1[(\alpha_2-\beta_2(1-s))\ln(1-\gamma_2)+\beta_2(1-s)\ln(1-\gamma_1)]+\delta_1 C_1\,. \end{array}\right.$$

Solving this system we obtain

$$A_1 = \frac{1-\delta_1(\alpha_2-\beta_2(1-s))}{\Delta_1}\,, \quad B_1 = \frac{\delta_1\beta_1 s}{\Delta_1}\,,$$

where

$$\Delta_1 = (1 - \delta_1\alpha_2 + \delta_1\beta_2(1-s))(1 - \delta_1\alpha_1 + \delta_1\beta_1 s) - \delta_1^2 s(1-s)\beta_1\beta_2 .$$

Similarly, for player 2 from (4.13) we receive

$$B_2 = \frac{1 - \delta_2(\alpha_1 - \beta_1 s)}{\Delta_2}, \quad A_2 = \frac{\delta_2\beta_2(1-s)}{\Delta_2},$$

where

$$\Delta_2 = (1 - \delta_2\alpha_2 + \delta_2\beta_2(1-s))(1 - \delta_2\alpha_1 + \delta_2\beta_1 s) - \delta_2^2 s(1-s)\beta_1\beta_2 .$$

To determine the optimal catch we maximize the right-hand side of (4.14)

$$\frac{1}{u_1} + \frac{\delta_1 A_1(-\alpha_1 + \beta_1 s) - \delta_1 B_1\beta_2(1-s)}{x - u_1} = 0$$

from where we obtain

$$u_1 = \frac{x\Delta_1}{1 - \delta_1\alpha_2 + \delta_1\beta_2(1-s)} = \gamma_1^* x .$$

Analogously, for player 2 we get

$$u_2 = \frac{y\Delta_2}{1 - \delta_2\alpha_1 + \delta_2\beta_1 s} = \gamma_2^* y .$$

Now we show that the catch of fish is feasible, i.e. $0 < \gamma_i^* < 1, i = 1, 2$.
For player 1 we get the condition

$$(1 - \delta_1\alpha_2)(\beta_1 s - \alpha_1) - \alpha_1\delta_1\beta_2(1-s) < 0$$

or

$$s < \frac{\alpha_1(1 - \delta_1\alpha_2) + \delta_1\alpha_1\beta_2}{\beta_1(1 - \delta_1\alpha_2) + \delta_1\alpha_1\beta_2} .$$

This inequality holds for all $0 \leq s \leq 1$ under our assumption $\alpha_1 > \beta_1$.
Similarly, for player 2 condition

$$s > \frac{(\beta_2 - \alpha_2)(1 - \delta_2\alpha_1)}{\beta_2(1 - \delta_2\alpha_1) + \delta_2\alpha_2\beta_1}$$

holds for all $0 \leq s \leq 1$ under our assumption $\alpha_2 > \beta_2$.

The steady-state stocks are derived from

$$\bar{x} = (\bar{x}(1-\gamma_1))^{\alpha_1 - \beta_1 s}(\bar{y}(1-\gamma_2))^{\beta_1 s}$$
$$\bar{y} = (\bar{y}(1-\gamma_2))^{\alpha_2 - \beta_2(1-s)}(\bar{x}(1-\gamma_1))^{\beta_2(1-s)}$$

with the solution

$$\bar{x} = (1-\gamma_2)^{\frac{\beta_1 s}{K}}(1-\gamma_1)^{\frac{(1-\alpha_2)(\alpha_1 - \beta_1 s) + \alpha_1\beta_2(1-s)}{K}},$$

$$\bar{y} = (1-\gamma_1)^{\frac{\beta_2(1-s)}{K}}(1-\gamma_2)^{\frac{(1-\alpha_1)(\alpha_2-\beta_2(1-s))+\alpha_2\beta_1 s}{K}},$$

where

$$K = (1-\alpha_1)(1-\alpha_2) + (1-\alpha_2)\beta_1 s + (1-\alpha_1)\beta_2(1-s)\,.$$

If we consider the case

$$\alpha_1 = \alpha_2 = \alpha,\ \beta_1 = \beta_2 = \beta,\ s = \frac{1}{2},$$

then

$$\gamma_1 = \gamma_2 = \gamma$$

and the steady-state is

$$\bar{x} = \bar{y} = (1-\gamma)^{\frac{\alpha}{1-\alpha}}\,.$$

Cooperative case. We continue our analysis with a cooperative case investigation. Under cooperation the object is to maximize the sum of the players' utilities. The corresponding problem is

$$\max_{u_{1t},u_{2t}} \sum_{t=0}^{\infty} \delta^t(\mu_1 \ln(u_{1t}) + \mu_2 \ln(u_{2t}))\,, \tag{4.15}$$

where $0 < \delta < 1$ – the common discount factor, $0 < \mu_1,\ \mu_2 < 1$ – weighting coefficients $(\mu_1 + \mu_2 = 1)$.

Theorem 4.5. *Cooperative equilibrium of the problem (4.10),(4.15) is*

$$u_1^d = \frac{\mu_1\Delta}{\mu_1(1-\delta\alpha_2) + \delta\beta_2(1-s)}x,\quad u_2^d = \frac{\mu_2\Delta}{\mu_2(1-\delta\alpha_1) + \delta\beta_1 s}y\,, \tag{4.16}$$

where

$$\Delta = (1-\delta\alpha_1)(1-\delta\alpha_2) + \delta\beta_2(1-s)(1-\delta\alpha_1) + \delta\beta_1 s(1-\delta\alpha_2)\,.$$

Proof. The proof is similar to the proof of Theorem 4.4. We only remark that the catch is feasible. For player 1 we get the condition

$$\mu_1[(1-\delta\alpha_2)(\beta_1 s - \alpha_1) - \alpha_1\delta\beta_2(1-s)] - \mu_2\delta\beta_2(1-s) < 0$$

or

$$s < \frac{\mu_1\alpha_1(1-\delta\alpha_2) + \delta\beta_2(\mu_1\alpha_1 + \mu_2)}{\mu_1\beta_1(1-\delta\alpha_2) + \delta\beta_2(\mu_1\alpha_1 + \mu_2)}\,.$$

This inequality holds for all $0 \le s \le 1$ under our assumption $\alpha_1 > \beta_1$.
Similarly, for player 2 condition

$$s > \frac{\mu_2(\beta_2 - \alpha_2)(1 - \delta\alpha_1)}{\mu_2\beta_2(1 - \delta\alpha_1) + \delta\beta_1(\mu_2\alpha_2 + \mu_1)}$$

holds for all $0 \le s \le 1$ under our assumption $\alpha_2 > \beta_2$.

Analogously to the noncooperative case, if we denote

$$\gamma_1^d = \frac{\mu_1\Delta}{\mu_1(1 - \delta\alpha_2) + \delta\beta_2(1 - s)}, \ \gamma_2^d = \frac{\mu_2\Delta}{\mu_2(1 - \delta\alpha_1) + \delta\beta_1 s}$$

then the steady-state stocks under cooperation are

$$\bar{x}^d = (1 - \gamma_2^d)^{\frac{\beta_1 s}{K}}(1 - \gamma_1^d)^{\frac{(1-\alpha_2)(\alpha_1 - \beta_1 s) + \alpha_1\beta_2(1-s)}{K}},$$

$$\bar{y}^d = (1 - \gamma_1^d)^{\frac{\beta_2(1-s)}{K}}(1 - \gamma_2^d)^{\frac{(1-\alpha_1)(\alpha_2 - \beta_2(1-s)) + \alpha_2\beta_1 s}{K}},$$

where

$$K = (1 - \alpha_1)(1 - \alpha_2) + (1 - \alpha_2)\beta_1 s + (1 - \alpha_1)\beta_2(1 - s).$$

If we consider the case

$$\alpha_1 = \alpha_2 = \alpha, \ \beta_1 = \beta_2 = \beta, \ s = \mu_1 = \mu_2 = \frac{1}{2},$$

then

$$\gamma_1^d = \gamma_2^d = \gamma^d$$

and the steady-state is

$$\bar{x}^d = \bar{y}^d = (1 - \gamma^d)^{\frac{\alpha}{1-\alpha}}.$$

For $\delta_1 = \delta_2 = \delta$ we can compare the stock in the noncooperative and cooperative cases and we conclude that

$$\gamma = \frac{(1 - \delta\alpha)(1 - \delta\alpha + \delta\beta)}{1 - \delta\alpha + 1/2\delta\beta} > \frac{(1 - \delta\alpha)(1 - \delta\alpha + \delta\beta)}{1 - \delta\alpha + \delta\beta} = \gamma^d$$

and

$$\bar{x} = \bar{y} < \bar{x}^d = \bar{y}^d.$$

This result shows that when there is an interaction between the regions, noncooperative behavior is inefficient due to overfishing.

Incentive equilibrium. Here we consider the incentive equilibrium as a method of keeping cooperation, moreover the center (referee) punishes the players for deviation.

As before $s^d = s$ – water area sharing in cooperative equilibrium.

Assume as above that, if the first player deviates the center decreases s, but if the second player deviates – increases s.

Theorem 4.6. *The incentive equilibrium of the problem (4.9),(4.10),(4.15) is*

$$\gamma_1(u_2) = \frac{\mu_1 \Delta_1^* x}{\mu_1(1-\delta\alpha_2) + \delta\beta_2(1-s_1^*)}\,, \quad \gamma_2(u_1) = \frac{\mu_2 \Delta_2^* y}{\mu_2(1-\delta\alpha_1) + \delta\beta_1 s_2^*}\,, \tag{4.17}$$

where

$$\Delta_1^* = (1-\delta\alpha_1)(1-\delta\alpha_2) + \delta\beta_2(1-s_1^*)(1-\delta\alpha_1) + \delta\beta_1 s_1^*(1-\delta\alpha_2)\,,$$

$$\Delta_2^* = (1-\delta\alpha_1)(1-\delta\alpha_2) + \delta\beta_2(1-s_2^*)(1-\delta\alpha_1) + \delta\beta_1 s_2^*(1-\delta\alpha_2)\,,$$

$$s_1^* = s^d + \eta_1(u_2 - u_2^d)\,, \quad s_2^* = s^d - \eta_2(u_1 - u_1^d)\,.$$

And the coefficients are

$$\eta_1 = \frac{\mu_1\beta_1 s^d}{\mu_2\beta_2(y-u_2^d)}\left[M_1 \ln\Big(\frac{x-u_1^d}{y-u_2^d}\Big) + \frac{(1-s^d)u_1^d}{(x-u_1^d)\Delta}\frac{(1-\delta\alpha_2)(\mu_1 N + \delta\beta_2 M_1)}{\mu_1(1-\delta\alpha_2)+\delta\beta_2(1-s^d)}\right]^{-1}, \tag{4.18}$$

$$\eta_2 = \frac{\mu_2\beta_2(1-s^d)}{\mu_1\beta_1(x-u_1^d)}\left[M_2 \ln\Big(\frac{y-u_2^d}{x-u_1^d}\Big) + \frac{s^d u_2^d}{(y-u_2^d)\Delta}\frac{(1-\delta\alpha_1)(-\mu_2 N + \delta\beta_1 M_2)}{\mu_2(1-\delta\alpha_1)+\delta\beta_1 s^d}\right]^{-1}, \tag{4.19}$$

where

$$M_i = 1 - \delta\alpha_i + \delta\beta_i\,,\ i = 1,2,$$

$$N = \delta\beta_1(1-\delta\alpha_2) - \delta\beta_2(1-\delta\alpha_1)\,.$$

Proof. Let us consider the second player's deviation

$$u_2 = u_2^d + \Delta\,.$$

We find the center's strategy in the following form

$$s^* = s^d + \eta_1(u_2 - u_2^d)\,.$$

Then the first player's punishment strategy is

$$\gamma_1(u_2) = \frac{\mu_1 x[(1-\delta\alpha_1)(1-\delta\alpha_2) + \delta\beta_2(1-s^*)(1-\delta\alpha_1)] + \delta\beta_1 s^*(1-\delta\alpha_2)}{\mu_1(1-\delta\alpha_2)+\delta\beta_2(1-s^*)}\,.$$

Next we should determine the coefficient η_1. For that we solve the problem for player 2 assuming that player 1 uses the punishment strategy, so

$$\max_{u_{2t}} \sum_{t=0}^{\infty} \delta^t \ln(u_{2t})\,. \tag{4.20}$$

And dynamics of the fishery becomes

$$\begin{cases} x_{t+1} = (x_t - \gamma_1(u_{2t}))^{\alpha_1}\left(\frac{y_t - u_{2t}}{x_t - \gamma_1(u_{2t})}\right)^{\beta_1 s^*}, \\ y_{t+1} = (y_t - u_{2t})^{\alpha_2}\left(\frac{x_t - \gamma_1(u_{2t})}{y_t - u_{2t}}\right)^{\beta_2(1-s^*)}, \end{cases} \tag{4.21}$$

where

$$s^* = s^d + \eta_{1t}(u_{2t} - u_{2t}^d).$$

For η_1 to be the incentive equilibrium the solution of the problem (4.20),(4.21) must be archived on u_2^d.

From this fact we find η_1, as an expression in (4.18).

Analogously if player 1 deviates we find the center's strategy in the following form

$$s^* = s^d - \eta_2(u_1 - u_1^d).$$

Then the second player's punishment strategy is

$$\gamma_2(u_1) = \frac{\mu_2 y[(1-\delta\alpha_1)(1-\delta\alpha_2) + \delta\beta_2(1-s^*)(1-\delta\alpha_1)] + \delta\beta_1 s^*(1-\delta\alpha_2)}{\mu_2(1-\delta\alpha_1) + \delta\beta_1 s^*}.$$

To determine the coefficient η_2 we solve the problem for player 1 assuming that player 2 uses the punishment strategy, so

$$\max_{u_{1t}} \sum_{t=0}^{\infty} \delta^t \ln(u_{1t}).$$

And dynamics of the fishery becomes

$$\begin{cases} x_{t+1} = (x_t - u_{1t})^{\alpha_1}\left(\frac{y_t - \gamma_2(u_{1t})}{x_t - u_{1t}}\right)^{\beta_1 s^*}, \\ y_{t+1} = (y_t - \gamma_2(u_{1t}))^{\alpha_2}\left(\frac{x_t - u_{1t}}{y_t - \gamma_2(u_{1t})}\right)^{\beta_2(1-s^*)}, \end{cases}$$

where

$$s^* = s^d - \eta_{2t}(u_{1t} - u_{1t}^d).$$

So we receive η_2, as an expression given in (4.19).

The economic feasibility of the center's cooperation maintenance is advantageous for the players who keep the agreement that was achieved at the beginning of the game. Thus, they cannot monitor the opponent actions that incur additional costs, and completely rely on the center.

Time-consistency. Let us consider the case when $\alpha_1 = \alpha_2 = \alpha$ and $\beta_1 = \beta_2 = \beta$. So, the fish populations grow according to the biological rule

$$\begin{cases} x_{t+1} = x_t^{\alpha}\left(\frac{y_t}{x_t}\right)^{\beta s}, \quad x_0 = x, \\ y_{t+1} = y_t^{\alpha}\left(\frac{x_t}{y_t}\right)^{\beta(1-s)}, \quad y_0 = y. \end{cases} \tag{4.22}$$

As before the players wish to maximize their total utilities over infinite time horizon:

$$J_1 = \sum_{t=0}^{\infty} \delta^t \ln(u_{1t}), \quad J_2 = \sum_{t=0}^{\infty} \delta^t \ln(u_{2t}), \tag{4.23}$$

where $0 \leq u_{1t} \leq x_t$, $0 \leq u_{2t} \leq y_t$ – countries' catch at a time t, $0 < \delta < 1$ – the common discount factor.

Now we consider the center as a player and his net revenue over infinite time horizon is

$$I = \sum_{t=0}^{\infty} \delta^t \ln(x_t y_t). \tag{4.24}$$

So the center wishes to maximize the total amount of fish in the reservoir.

For presented model we find Shapley value and time-consistent imputation distribution procedure.

Theorem 4.7. *If* $x_0 = y_0$ *then the Shapley value for the problem (4.22)-(4.24) is*

$$(\xi_1, \xi_2, \xi_s),$$

where

$$\xi_1 = \frac{2-\mu_1}{2(1-\alpha\delta)} \ln x + \frac{1}{6(1-\delta)}(2C_1 + 2C_{1,2,s} - 2C_{2,s} + C_{1,2} + C_{1,s} - C_2 - C_s),$$

$$\xi_2 = \frac{1+\mu_1}{2(1-\alpha\delta)} \ln x + \frac{1}{6(1-\delta)}(-C_1 + 2C_{1,2,s} + C_{2,s} + C_{1,2} - 2C_{1,s} + 2C_2 - C_s),$$

$$\xi_s = \frac{3}{2(1-\alpha\delta)} \ln x + \frac{1}{6(1-\delta)}(-C_1 + 2C_{1,2,s} + C_{2,s} - 2C_{1,2} + C_{1,s} - C_2 + 2C_s),$$

and

$C_1 = C_2 = \ln(\frac{\Delta}{1-a+0.5b}) + \frac{a}{1-a}\ln(1 - \frac{\Delta}{1-a+0.5b})$, $C_s = \frac{2a}{(1-a)}\ln(1 - \frac{\Delta}{1-a+0.5b})$,

$C_{1,2,s} = \ln(\frac{1-a}{1+2a}) + \frac{3a}{1-a}\ln(1 - \frac{1-a}{1+2a})$, $C_{1,2} = \ln(1-a) + \frac{a}{1-a}\ln(a)$,

$C_{1,s} = \mu_1 \ln(\frac{2\Delta}{2-\mu_1 a(1-a+b)-(a-b)(a+1)}) + \frac{(2-\mu_1)a}{1-a}\ln(1 - \frac{2\Delta}{2-\mu_1 a(1-a+b)-(a-b)(a+1)})$,

$C_{2,s} = (1-\mu_1)\ln(\frac{2\Delta}{2+\mu_1 a(1-a+b)-2a+b}) + \frac{(1+\mu_1)a}{1-a}\ln(1 - \frac{2\Delta}{2+\mu_1 a(1-a+b)-2a+b})$,

where $a = \alpha\delta$, $b = \beta\delta$, $\Delta = (1-a)(1-a+b)$.

Proof. All the following proof is right if $x_0 \neq y_0$, but the coefficients and center's control are more complicated. So we give here proof only for simplest case.

For solving this problem we consider all possible coalitions, find optimal players' strategies. For that we use the same approach as in Theorems 4.4 and 4.5.

We find player's strategies in the following form

$$u_{1t} = \gamma_1 x_t, \quad u_{2t} = \gamma_2 y_t.$$

For all coalitions we receive that the center's optimal control is

$$s : \frac{x_t(1-\gamma_1)}{(1-\gamma_2)y_t} = 1$$

and below we give the expression for s only for the case $x_0 = y_0$.
Namely, we received:

1. Coalitions $\{1\}, \{2\}, \{s\}$.

$$s = \frac{1}{2}$$

$$\gamma_1 = \gamma_2 = \frac{\Delta}{1 - \alpha\delta + 0.5\beta\delta},$$

where $\Delta = (1-\alpha\delta)(1-\alpha\delta+\delta\beta)$.

2. Grand coalition $\{1, 2, s\}$.

$$s = \frac{\alpha(1-\alpha\delta+\delta\beta)(1-2\mu_1) + \beta(2-\mu_1)}{3\beta}$$

$$\gamma_1 = \gamma_2 = \frac{1-\alpha\delta}{1+2\alpha\delta}.$$

3. Coalitions $\{1, 2\}, \{s\}$.

$$s = 1 - \mu_1$$

$$\gamma_1 = \gamma_2 = 1 - \alpha\delta.$$

4. Coalitions $\{1, s\}, \{2\}$.

$$s = \frac{-\alpha\mu_1(1-\alpha\delta+\delta\beta) + \beta(1+\alpha\delta) + \alpha(1-\alpha\delta)}{2\beta}$$

$$\gamma_1 = \gamma_2 = \frac{2\Delta}{2 - \mu_1\delta\alpha(1-\alpha\delta+\delta\beta) - \delta(\alpha-\beta)(\alpha\delta+1)}.$$

5. Coalitions $\{2, s\}, \{1\}$.

$$s = \frac{-\alpha\mu_1(1-\alpha\delta+\delta\beta) + \beta}{2\beta}$$

$$\gamma_1 = \gamma_2 = \frac{2\Delta}{2 + \mu_1\delta\alpha(1-\alpha\delta+\delta\beta) - \delta(2\alpha-\beta)}.$$

As you can notice in the considered case $x_t = y_t$ for all possible coalitions, because $\frac{x_{t+1}}{y_{t+1}} = (\frac{x_t}{y_t})^{\alpha-\beta}$ and $x_0 = y_0$.

So if in general we find a profit in the form

$$V_{\{K\}} = A_K \ln x + B_K \ln y + C_K$$

for all coalitions K, under our assumption ($x_0 = y_0$) the profit is

$$V_{\{K\}} = (A_K + B_K) \ln x + C_K \,.$$

And we give it for all possible coalitions

$$\begin{array}{l} V_{\{1\}} = V_{\{2\}} = \frac{1}{1-\alpha\delta} \ln x + \frac{1}{1-\delta} C_1 \,, \quad V_{\{s\}} = \frac{2}{1-\alpha\delta} \ln x + \frac{1}{1-\delta} C_s \,, \\ V_{\{1,2,s\}} = \frac{3}{1-\alpha\delta} \ln x + \frac{1}{1-\delta} C_{1,2,s} \,, \quad V_{\{1,2\}} = \frac{1}{1-\alpha\delta} \ln x + \frac{1}{1-\delta} C_{1,2} \,, \\ V_{\{1,s\}} = \frac{2-\mu_1}{1-\alpha\delta} \ln x + \frac{1}{1-\delta} C_{1,s} \,, \quad V_{\{2,s\}} = \frac{1+\mu_1}{1-\alpha\delta} \ln x + \frac{1}{1-\delta} C_{2,s} \,, \end{array} \tag{4.25}$$

where coefficients $C_{i,j}$ are given in the statement.

From (4.25) we obtain the Shapley value.

The dynamics under cooperation will be

$$x_{t+1} = x_t^{\alpha} (1 - \gamma)^{\alpha} = x_t^{\alpha} (\frac{3\alpha\delta}{1 + 2\alpha\delta})^{\alpha}$$

which leads to

$$x_t = x_0^{\alpha^t} (\frac{3\alpha\delta}{1 + 2\alpha\delta})^{\sum\limits_{j=1}^{t} \alpha^j} \,.$$

Theorem 4.8. *The vector $\beta(t) = (\beta_1(t), \beta_2(t), \beta_s(t))$ where*

$$\beta_i(t) = \xi_i(t) - \delta \xi_i(t+1) \,, i = 1, 2, s$$

is time-consistent imputation distribution procedure.

Proof. Let's prove that this vector is imputation distribution procedure, from (2.6) we need to check

$$\sum_{t=0}^{\infty} \delta^t \beta_i(t) = \sum_{t=0}^{\infty} \delta^t \xi_i(t) - \sum_{t=0}^{\infty} \delta^{t+1} \xi_i(t+1) = \xi_i(0) \,.$$

Let's prove that this vector is time-consistent imputation distribution procedure, from (2.8), we consider

$$\sum_{\tau=0}^{t-1} \delta^{\tau} \beta_i(\tau) + \delta^t \xi_i(t) = \sum_{\tau=0}^{t-1} \delta^{\tau} \xi_i(\tau) - \sum_{\tau=0}^{t-1} \delta^{\tau+1} \xi_i(\tau+1) + \delta^t \xi_i(t) = \xi_i(0) \,.$$

The irrational-behavior-proffnes condition in our model has the form

$$\xi_i(0) - \xi_i(t) \delta^t \geq V_{\{i\}}(0) - \delta^t V_{\{i\}}(t) \,.$$

And for our model each step rational behavior condition has the form

$$\xi_i(t) - \delta\xi_i(t+1) \geq V_{\{i\}}(t) - \delta V_{\{i\}}(t+1) .$$

As you can notice the each step rational behavior condition yields the Yeung's condition, so we prove the modified one.

For the case $x_t = y_t$ it takes the form:

For player 1

$$-\frac{\mu_1}{2}\ln(x_t) + \frac{1}{6}[\frac{3\mu_1\alpha\delta}{1-\alpha\delta}\ln(\frac{3\alpha\delta}{1+2\alpha\delta}) - 5C_1 + 2C_{1,2,s} - 2C_{2,s} + C_{1,2} + C_{1,s} - C_s] \geq 0 .$$

For player 2

$$\frac{\mu_1 - 1}{2}\ln(x_t) + \frac{1}{6}[\frac{3(1-\mu_1)\alpha\delta}{1-\alpha\delta}\ln(\frac{3\alpha\delta}{1+2\alpha\delta}) - 5C_1 + 2C_{1,2,s} + C_{2,s} + C_{1,2} - 2C_{1,s} - C_s] \geq 0 .$$

For the center

$$-\frac{1}{2}\ln(x_t) + \frac{1}{6}[\frac{3\alpha\delta}{1-\alpha\delta}\ln(\frac{3\alpha\delta}{1+2\alpha\delta}) - 2C_1 + 2C_{1,2,s} + C_{2,s} - 2C_{1,2} + C_{1,s} - 4C_s] \geq 0 .$$

It is not difficult to verify that these conditions hold. For example, let us consider the first condition and prove that the expression in square brackets is positive.

$$\begin{aligned} D &= \frac{3\mu_1\alpha\delta}{1-\alpha\delta}\ln(\frac{3\alpha\delta}{1+2\alpha\delta}) - 5C_1 + 2C_{1,2,s} - 2C_{2,s} + C_{1,2} + C_{1,s} - C_s = \\ &= \frac{3\mu_1\alpha\delta}{1-\alpha\delta}\ln(\frac{3\alpha\delta}{1+2\alpha\delta}) - 5C_1 + 3C_{1,2,s} + (C_{1,s} - 2C_{2,s}) - C_s - \\ &- \frac{3\alpha\delta}{1-\alpha\delta}\ln(3\alpha\delta) + \frac{1+2\alpha\delta}{1-\alpha\delta}\ln(1+2\alpha\delta) + \frac{\alpha\delta}{1-\alpha\delta}\ln(\alpha\delta) . \end{aligned}$$

Note that this function is decreasing on μ_1, so we need to verify that $D \geq 0$ when $\mu_1 = 1$. Also see that $C_{1,s} - 2C_{2,s} > 0$. Then

$$\begin{aligned} D &> 3C_{1,2,s} - 5C_1 - C_s + \ln(1+2\alpha\delta) + \frac{\alpha\delta}{1-\alpha\delta}\ln(\alpha\delta) = \\ &= \Big(3C_{1,2,s} - 5C_1 - \frac{3\alpha\delta}{2(1-\alpha\delta)}\ln(1 - \frac{\Delta}{1-\alpha\delta+(\beta\delta)/2})\Big) + \\ &+ \Big(\ln(1+2\alpha\delta) + \frac{\alpha\delta}{1-\alpha\delta}\ln(\alpha\delta) - \frac{\alpha\delta}{2(1-\alpha\delta)}\ln(1 - \frac{\Delta}{1-\alpha\delta+(\beta\delta)/2})\Big) , \end{aligned}$$

where $\Delta = (1-\alpha\delta)(1-\alpha\delta+\beta\delta)$.

And note that the expressions in the first and the second brackets are positive.

To complete the proof of the first condition recall that

$$\ln(x_t) < 0 .$$

4.3. Modelling

We consider this model which regulates the player's cooperative behavior for the next set of parameters

$$\alpha_1 = \alpha_2 = 0.4\,, \quad \rho = 0.1\,, \quad s^d = 0.5$$
$$\beta_1 = \beta_2 = 0.3\,, \quad \mu_1 = 0.55\,, \quad \mu_2 = 0.45\,.$$

These parameters are typical for the fish species in Karelian lakes (Sterligova et al., 2002; Mazalov & Rettieva, 2005, 2006).

First we compare the size of population in cooperative and noncooperative cases. Fig. 9 and 10 show population's dynamics in the regions in the case of Nash equilibrium (dark line) and in the case of cooperation (bright line). As you can see sizes of populations under cooperative behavior are larger than in the case of noncooperation. This result shows that noncooperative behavior is inefficient due to overfishing

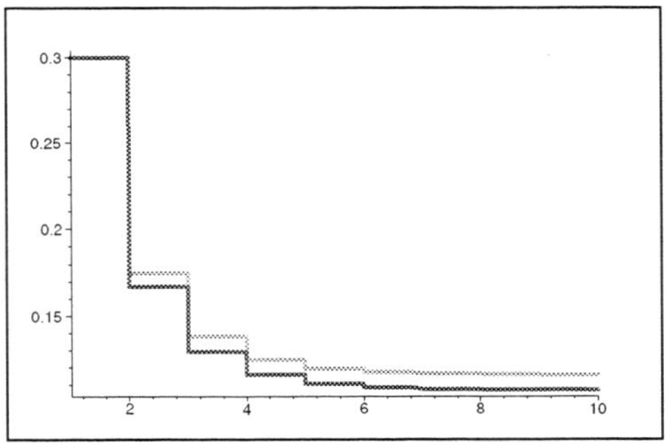

Figure 9. Size of population x_t.

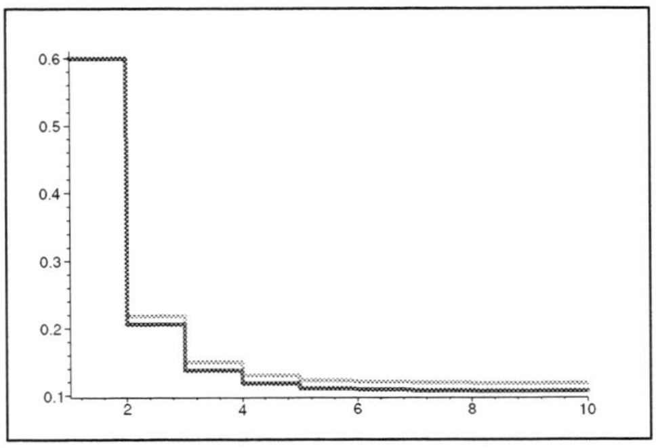

Figure 10. Size of population y_t.

Incentive equilibrium. The initial size of the population $x = 0.3$ and $y = 0.6$. The number of steps is 12. The time when the second player deviates is $t_0 = 5$ and the size of a deviation is $\Delta = 0.0001$.

In the figures, you can see the difference in parameters in the case of cooperative behavior (dark line) and in the case of the deviation (bright line). For better clarity of the difference we demonstrate the figures on the time interval $[4, 10]$. On Fig. 11 you can see x_t and on Fig. 12 – y_t. Fig. 13 and 14 present the first and the second player's catch, respectively. We notice that player 2 increases his catch on time interval $[5, 6]$ and decreases after, because of the punishment. The first player's catch decreases on time interval [5,6] and increases after.

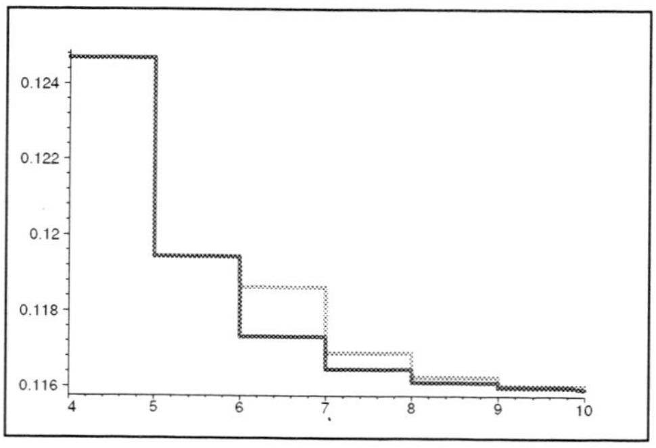

Figure 11. Size of population x_t.

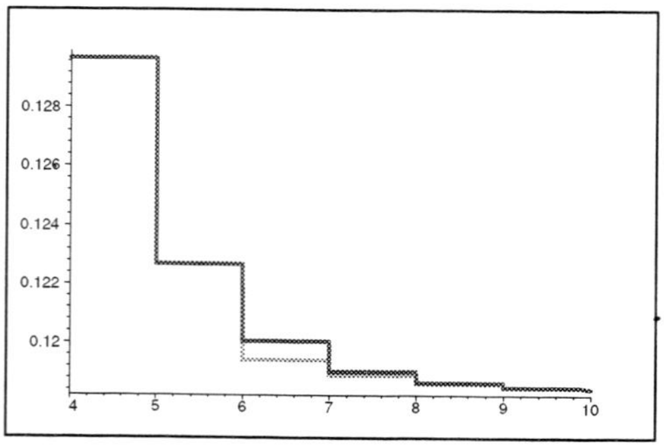

Figure 12. Size of population y_t.

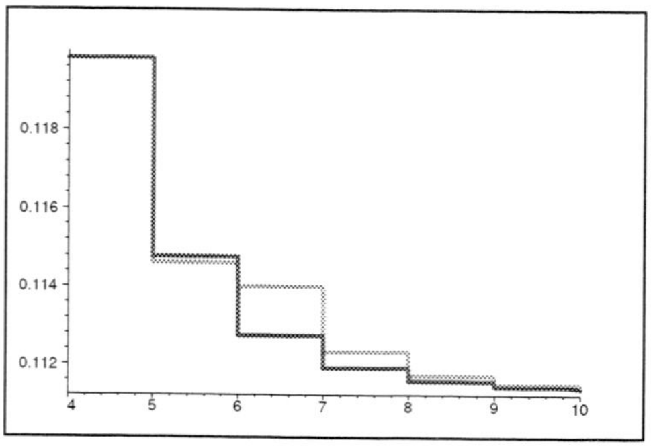

Figure 13. Catch of player 1.

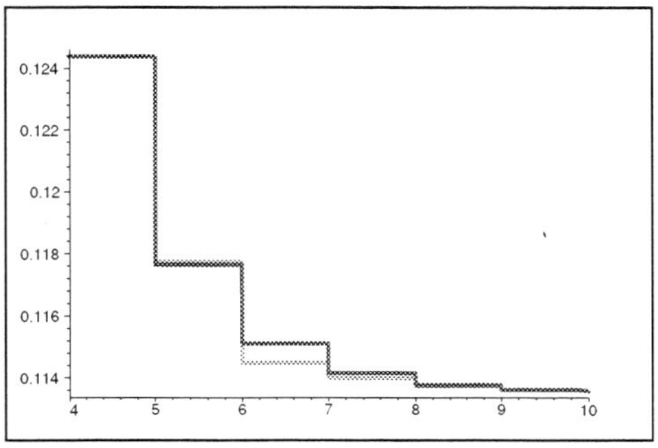

Figure 14. Catch of player 2.

Time-consistent imputation distribution procedure. The initial size of the population $x = y = 0.5$. The number of steps is 10.

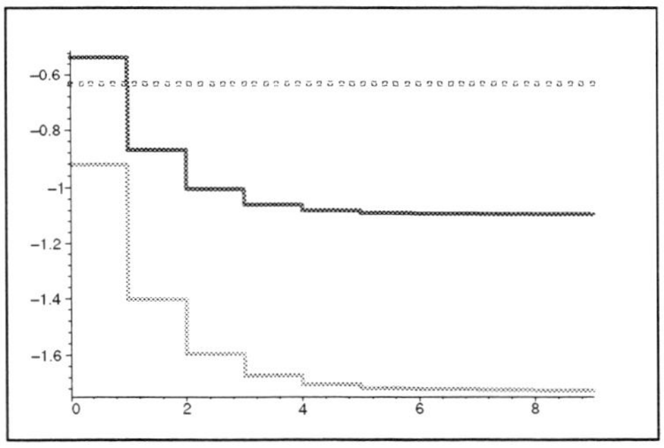

Figure 15. Distribution procedure for player 1.

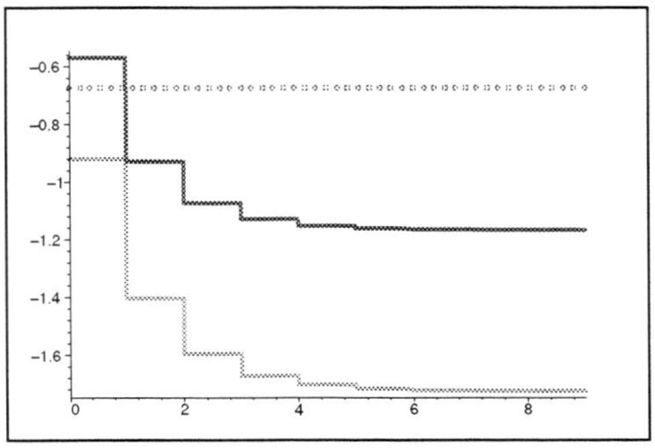

Figure 16. Distribution procedure for player 2.

In the Fig. 15-17 you can see time-consistent imputation distribution procedure ($\beta_i(t)$) for player i (dark line), player's i Nash profit – $V_{\{i\}}$ (bright line) and his Shapley value – $\xi_i(0)$ (dotted line).

You can notice that the distribution procedure is greater than profit in noncooperative

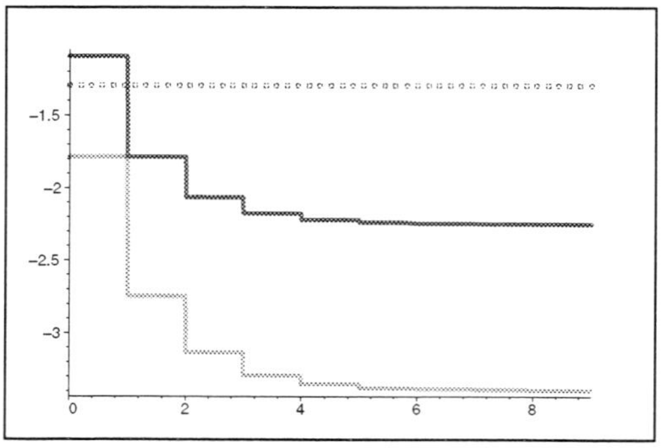

Figure 17. Distribution procedure for the center.

case at every time moment. Figures shows how to distribute the cooperation gain (Shapley value) along the game path.

5. Conclusion

In the chapter we introduce the results of continuous and discrete-time bioresource management problem investigation. The Nash, cooperative and incentive equilibria are derived for different growth rules and players' payoff functions.

We consider two schemes to maintain the cooperative behavior: cooperative incentive equilibrium and time-consistent imputation distribution procedure.

In the context of cooperative incentive equilibrium we investigate the case when the center punishes defaulter players in contrast to traditional scheme of cooperation maintenance. The center's strategy here is the territory sharing. The player who breaks the agreement achieved at the beginning of the game is punished by gradually decreasing the harvesting territory. The economic feasibility of center's cooperation maintenance is advantage for the players who keep agreement achieved at the beginning of the game. Thus they can not monitoring the opponent actions that incur additional costs, and completely rely on the center. So, in a case when players control each other's behavior, when the second player deviates the first player is compelled to increase his fishing efforts too, i.e. to incur additional cost on large number of ship's operation. In the case when the center punishes deviating players, the honest player reduces his fishing efforts conversely, but his catch increases that is connected with catch territory change. Thus, he gets bigger profit with smaller expenses for ship's operation.

In the context of time-consistency we construct imputation set and time-consistent imputation distribution procedure. We present a new condition which stimulates cooperative behavior and is easier to verify than the well-known Yeung's condition. This condition offers an incentive to player to maintain cooperation because at every step she gains more from cooperation than from noncooperative behavior.

Acknowledgements

The work was supported by Russian Fund for Basic Research, projects 06-01-00128a, 08-01-98801-r-sever-a and 10-01-00089-a.

References

[1] Abakumov, A.I.; Kol'ev, N.V.; Maximenko, V.P.; Gorr, S.V. *Ichthyology questions*. 1994, 34(3), 400–407 (in russian).

[2] Ehtamo, H.; Hamalainen, R. P. *J. of Econ. Dynamic and Control*. 1993, 17, 659–678.

[3] Fisher, R.D.; Mirman, L.J. *J. of Environmental Economics and Management*. 1996, 30, 34–42.

[4] Haurie, A. *J. of Optimiz. Theory and Appl*. 1976, 18, 31–39.

[5] Levhari, D.; Mirman, L.J. *The Bell J. of Economic*. 1980, 11(1), 322–334.

[6] Mazalov, V.V.; Rettieva, A.N. *Int. J. of Mathematics, Game Theory and Algebra*. 2005, 15(3), 243–256.

[7] Mazalov, V. V.; Rettieva, A. N. *Mathem. modelling*. 2006, 18(5), 73–90.(in russian)

[8] Mazalov, V.V.; Rettieva, A.N. *Game Theory and Applications*. 2008, 13, 101–110.

[9] Mazalov, V.V.; Rettieva, A.N. *Doklady Math*. 2008, 78(3), 320–322.

[10] Mazalov, V.V.; Rettieva, A.N. *Doklady Math*. 2010 (in publishing).

[11] Mazalov, V.V.; Rettieva, A.N. *Ecological Modelling*. 2010, 221, 1545–1553.

[12] Nowak, A. *Economics Bulletin*. 2006, *17(2)*, 1–10.

[13] Osborn, D.K. *American Economic Review*. 1976, 66, 835–844.

[14] Petrosjan, L.A. // *Viestnik LGU*. 1977, 19, 46–52.

[15] Petrosjan, L.A.; Danilov, N.N. *Cooperative differential games and their applications*. Izd Tomsk. Univ.: Tomsk, 1982.

[16] Petrosjan, L.A. *Differential game of pursuit*. World Scientific: Singapore, 1993.

[17] Petrosjan, L.A.; Zaccour, G. *J. Econ. Dyn. and Control*. 2003, 7, 381–398.

[18] Petrosjan, L.A.; Zenkevich, N.A. *Game theory*. World Scientific: Singapore, 1996.

[19] Pontryagin, L.S; Boltyanskii, V.G. and al. *Mathematical theory of optimal processes*. Nauka: Moscow, 1976.

[20] Sterligova, O.P.; Pavlov, V.N.; Ilmast, N.V. *The ecosystem of Lake Syamozero: biological regime and use*. Petrozavodsk. 2002 (in russian).

[21] Yeung, D.W.K. *Game Theory Rev*. 2006, 8(4), 739–744.

INDEX

A

B

C

D

G

H

I

N

O

P

T

U

V

W

Y

Z